Distributed Energy Storage Devices in Smart Grids

Distributed Energy Storage Devices in Smart Grids

Special Issue Editors

Guido Carpinelli
Pasquale De Falco
Fabio Mottola

MDPI • Basel • Beijing • Wuhan • Barcelona • Belgrade • Manchester • Tokyo • Cluj • Tianjin

Special Issue Editors

Guido Carpinelli
University of Napoli Federico II
Italy

Pasquale De Falco
University of Napoli Parthenope
Italy

Fabio Mottola
University of Napoli Federico II
Italy

Editorial Office
MDPI
St. Alban-Anlage 66
4052 Basel, Switzerland

This is a reprint of articles from the Special Issue published online in the open access journal *Energies* (ISSN 1996-1073) (available at: https://www.mdpi.com/journal/energies/special_issues/distributed_energy_storage_devices_smart_grids).

For citation purposes, cite each article independently as indicated on the article page online and as indicated below:

LastName, A.A.; LastName, B.B.; LastName, C.C. Article Title. *Journal Name* **Year**, *Article Number*, Page Range.

ISBN 978-3-03928-434-4 (Pbk)
ISBN 978-3-03928-435-1 (PDF)

© 2020 by the authors. Articles in this book are Open Access and distributed under the Creative Commons Attribution (CC BY) license, which allows users to download, copy and build upon published articles, as long as the author and publisher are properly credited, which ensures maximum dissemination and a wider impact of our publications.

The book as a whole is distributed by MDPI under the terms and conditions of the Creative Commons license CC BY-NC-ND.

Contents

About the Special Issue Editors vii

Preface to "Distributed Energy Storage Devices in Smart Grids" ix

Hyung Tae Kim, Young Gyu Jin and Yong Tae Yoon
An Economic Analysis of Load Leveling with Battery Energy Storage Systems (BESS) in an Electricity Market Environment: The Korean Case
Reprinted from: *Energies* **2019**, *12*, 1608, doi:10.3390/en12091608 1

Mayank Jha, Frede Blaabjerg, Mohammed Ali Khan, Varaha Satya Bharath Kurukuru and Ahteshamul Haque
Intelligent Control of Converter for Electric Vehicles Charging Station
Reprinted from: *Energies* **2019**, *12*, 2334, doi:10.3390/en12122334 17

Davide De Simone and Luigi Piegari
Integration of Stationary Batteries for Fast Charge EV Charging Stations
Reprinted from: *Energies* **2019**, *12*, 4638, doi:10.3390/en12244638 43

Andrea Mazza, Hamidreza Mirtaheri, Gianfranco Chicco, Angela Russo and Maurizio Fantino
Location and Sizing of Battery Energy Storage Units in Low Voltage Distribution Networks
Reprinted from: *Energies* **2020**, *13*, 52, doi:10.3390/en13010052 55

Fabio Mottola, Daniela Proto, Pietro Varilone and Paola Verde
Planning of Distributed Energy Storage Systems in μGrids Accounting for Voltage Dips
Reprinted from: *Energies* **2020**, *13*, 401, doi:10.3390/en13020401 75

Nayeem Chowdhury, Fabrizio Pilo and Giuditta Pisano
Optimal Energy Storage System Positioning and Sizing with Robust Optimization
Reprinted from: *Energies* **2020**, *13*, 512, doi:10.3390/en13030512 95

Abbas Marini, Luigi Piegari, S-Saeedallah Mortazavi and Mohammad-S Ghazizadeh
Coordinated Operation of Energy Storage Systems for Distributed Harmonic Compensation in Microgrids
Reprinted from: *Energies* **2020**, *13*, 771, doi:10.3390/en13030771 115

About the Special Issue Editors

Guido Carpinelli (Full Professor) is currently with the Department of Electrical Engineering and Information Technology, University of Napoli Federico II, Italy. His research interests mainly include power quality, power system analysis, and energy forecasting. He has served as Editor and Technical Committee Member for several international journals and international conferences. He has co-authored more than 200 papers in relevant journals and conference proceedings, three books in Italian, three books in English, and one book in Chinese. He was Guest Editor of two Special Issues in MDPI journals.

Pasquale De Falco (Post-doc researcher) received a Ph.D. in Information Technology and Electrical Engineering from the University of Napoli Federico II, Italy in 2018. He is currently with the Department of Engineering, University of Napoli Parthenope, Italy. His research interests mainly include energy forecasting, energy data analytics, and probabilistic methods applied to power systems. He serves as Editor of two MDPI journals (*Electronics and Forecasting*) and he was Guest Editor of four Special Issues in MDPI journals.

Fabio Mottola (Assistant Professor) received a Ph.D. in Electrical Engineering from the University of Napoli Federico II, Italy in 2008. He currently is with the Department of Electrical Engineering and Information Technology, University of Napoli Federico II, Italy. His research interests mainly include the planning and operation of electrical energy systems and lightning effects on power lines. He is a Senior Member of the Institute of Electrical and Electronics Engineers.

Preface to "Distributed Energy Storage Devices in Smart Grids"

Energy storage systems have been recognized as viable solutions for implementing the smart grid paradigm, but have created challenges for load levelling, integrating renewable and intermittent sources, voltage and frequency regulation, grid resiliency, improving power quality and reliability, reducing energy import during peak demand periods, and so on. In particular, distributed energy storage addresses a wide range of the above potential issues, and it is gaining attention from customers, utilities, and regulators.

Distributed energy storage has considerable potential for reducing costs and improving the quality of electric services. However, installation costs and lifespan are the main drawbacks to the wide diffusion of this technology. In this context, a serious challenge is the adoption of new techniques and strategies for the optimal planning, control, and management of grids that include distributed-energy storage devices. Regulatory guidance and proactive policies are urgently needed to ensure a smooth rollout of this technology.

This collection of recent contributions addresses the development of methodologies applied to the integration of distributed energy storage devices in smart power systems. Several areas of research, such as optimal siting and sizing of energy storage systems, adaption of energy storage systems to load leveling and harmonic compensation, integration of electric vehicles and optimal control systems, are investigated in the contributions collected in this book, including:

- An economic analysis of load leveling with battery energy storage systems in an electricity market from the perspective of a utility company and/or a government agency (Kim, H.; Jin, Y.; Yoon, Y. An Economic Analysis of Load Leveling with Battery Energy Storage Systems (BESS) in an Electricity Market Environment: The Korean Case. *Energies* 2019, 12(9), 1608);

- A proposal of a multiport converter for integrating photovoltaics, electric vehicle charging docks, and energy storage devices with the grid system (Jha, M.; Blaabjerg, F.; Khan, M.; Bharath Kurukuru, V.; Haque, A. Intelligent Control of Converter for Electric Vehicles Charging Station. *Energies* 2019, 12(12), 2334);

- The configuration and control of a converter able to integrate batteries in a fast electric vehicle charging station by decoupling the grid power from vehicle power using several auxiliary battery modules (De Simone, D.; Piegari, L. Integration of Stationary Batteries for Fast Charge EV Charging Stations. *Energies* 2019, 12(24), 4638);

- A method that fully supports battery energy storage system location and sizing, considering the time-variable generation and demand patterns and by considering both technical and economic aspects (Mazza, A.; Mirtaheri, H.; Chicco, G.; Russo, A.; Fantino, M. Location and Sizing of Battery Energy Storage Units in Low Voltage Distribution Networks. *Energies* 2020, 13(1), 52);

- A multi-step procedure for the optimal planning of the electrical energy storage systems in the microgrids aimed at minimizing the total costs while considering the compensation of the voltage dips (Mottola, F.; Proto, D.; Varilone, P.; Verde, P. Planning of Distributed Energy Storage Systems in μGrids Accounting for Voltage Dips. Energies 2020, 13(2), 401).

- A methodology for energy storage placement in the distribution networks, in which robust optimization accommodates system uncertainty (Chowdhury, N.; Pilo, F.; Pisano, G. Optimal Energy Storage System Positioning and Sizing with Robust Optimization. *Energies* 2020, 13(3), 512).

- An optimization model to operate energy storage systems as coordinated active harmonic filters for distributed compensation (Marini, A.; Piegari, L.; Mortazavi, S.; Ghazizadeh, M. Coordinated Operation of Energy Storage Systems for Distributed Harmonic Compensation in Microgrids. *Energies* 2020, 13(3), 771).

This book is relevant for energy practitioners involved in smart grid planning, management, and control. It also represents a good starting point for young researchers who want to discover the potential and up-to-date areas of research related to the services provided by distributed energy storage systems in smart grid context. Eventually, the contents of the book may be used by system operators to address the planning of future installations of energy storage system to boost the performance of networks.

Guido Carpinelli, Pasquale De Falco, Fabio Mottola
Special Issue Editors

Article

An Economic Analysis of Load Leveling with Battery Energy Storage Systems (BESS) in an Electricity Market Environment: The Korean Case

Hyung Tae Kim [1], Young Gyu Jin [2,*] and Yong Tae Yoon [1]

1 Department of Electrical and Computer Engineering, Seoul National University, 1 Gwanak-ro, Gwanak-gu, Seoul 08826, Korea; laputa11@snu.ac.kr (H.T.K.); ytyoon@snu.ac.kr (Y.T.Y.)
2 Department of Electrical Engineering, Jeju National University, 102 Jejudaehak-ro, Jeju-si, Jeju 63243, Korea
* Correspondence: ygjin93@jejunu.ac.kr; Tel.: +82-64-754-3677

Received: 22 March 2019; Accepted: 23 April 2019; Published: 27 April 2019

Abstract: The capacity of battery energy storage systems (BESS) is expected to increase for power system applications. However, as the cost of BESS is high, economic feasibility must be considered when using BESS in grid applications. Load leveling with BESS is one such application for which the economic implications have been analyzed in the literature. However, these studies do not sufficiently consider the fact that the leveled loads will lead to a change in electricity prices, thereby modifying charging/discharging operations of BESS. Additionally, in a competitive electricity market, electricity prices are not determined by the generator cost functions. Market participants' strategic decisions also affect prices. Therefore, we conducted an economic analysis of load leveling with BESS in an electricity market from the perspective of a utility company and/or a government agency. In our analysis of the Korean market, we examine whether the leveled loads necessarily lead to economic benefits. Load leveling performance and the associated economic benefit are quantitatively analyzed for varying sizes of BESS. Further, the policy implications related to using BESS are derived from the analysis results.

Keywords: battery energy storage systems (BESS); economic analysis; load leveling; electricity market

1. Introduction

The urgent environmental need to reduce greenhouse gas emissions has led to the integration of renewable energy sources (RESs) into the power grid. In 2018, the global cumulative installed capacity of RESs such as wind power and solar photovoltaic (PV) generation amounted to 539,123 MW [1] and 402.5 GW [2], respectively. However, RESs have disadvantages in terms of variability, intermittency, and uncertainty of electricity generation [3]. Therefore, energy storage systems (ESS), particularly battery ESS (BESS), have recently been attracting significant interest in the electricity industry as an effective means of mitigating the disadvantages of RESs [4,5].

The installed BESS capacity for power system applications is expected to increase consistently, not only for utility-scale applications but also for distributed small-scale applications, whereby a large number of BESS are expected to be connected to PV systems and behind the meter [6,7]. This trend of increasing BESS usage is likely to be accelerated by the widespread adoption of electric vehicles (EVs), supported by improvements in vehicle-to-grid and grid-to-vehicle technologies [8]. Furthermore, cost reductions in BESS technology are widely expected to become a key determinant of their increased installation. For instance, it is expected that the cost of lithium-ion BESS will decrease significantly, namely by 54–61%, between 2016 and 2030 [6], or by 52% between 2018 and 2040 [9]. For the latter period, it is estimated that USD 620 billion in investment in the vehicle and the electricity sectors will ensue [9].

Despite the positive forecasts regarding the cost of BESS, currently, costs are still so high that the economic feasibility of their installation is not always given [10]. For this reason, several studies have dealt with the economic aspects of BESS [11–38]. For instance, studies examine the BESS size and the associated charging/discharging strategy for optimal usage, either by minimizing total cost or maximizing the net present value (NPV) of grid-connected renewable energy systems [11–15]. Further, for a given BESS capacity, studies examine the operating strategy (i.e., the charging/discharging schedule) is determined to minimize the cost or maximize the net benefit under a dynamic pricing scheme [16–18]. Other researchers consider how a bidding strategy for BESS in an electricity market environment can also be developed to increase profit [19,20]. The economic aspects of BESS have also been considered for EV charging station planning [21–23]. Finally, economic considerations are important in the process of designing and operating a microgrid with ESS [24–26], where the methods based on particle swarm optimization have been used to determine an optimal size and/or an optimal operational strategy of ESS in a microgrid. Particular consideration is given to the integration of RESs into the microgrid in [24–26].

Economic feasibility is also an important factor to be considered for the various grid applications of BESS [5,27]. This study focuses on the application of load leveling and analyzes its economic benefits. Peak shaving using the discharging operation of BESS (when electricity demand is particularly high) is inevitably followed by the charging operation during off-peak periods. Therefore, the analysis of load leveling with BESS in this study includes peak shaving. Previous studies dealing with the economic aspects of load leveling with BESS obtained the following results: It was found that load leveling with BESS achieves an increase in the cost savings of thermal units [28–30]. Further, research suggests that BESS can be used to maximize the NPV of a renewable energy system and a substation [31,32]. Evidence also suggests that the leveled loads with BESS may provide economic benefits by reducing the required capacity of other equipment, such as generators and transformers, and by reducing the frequency with which investment in such network equipment is needed [33]. Finally, it has been shown that an economic benefit can also be obtained by peak shaving or peak reduction with BESS under a dynamic electricity pricing scheme [34–36]. Since the accurate forecasts of generation and load are essential to improving the performance of peak shaving, a method for peak shaving combined with a forecasting method based on the adaptive neuro-fuzzy inference system is presented in [37].

Customers of electricity providers, particularly the industrial customers, pay the electricity bill, which is comprised of the consumed energy cost and the peak demand surcharge. Thus, the economic benefit of load leveling or peak reduction with BESS is obvious to customers. Since the buying cost of electricity is a major cost component of a utility company, the provider's economic benefit with BESS corresponds to the reduction in the purchasing cost. Further, this economic benefit is relevant for the government agencies which are responsible for the policies of the electricity industry because the government permits billing rates to customers considering the cost of a utility. It can induce a utility to lower the bill if the cost can be reduced by load leveling with BESS. This study considers the perspective of a utility company and/or government agency. Hence, the problem to address in this study can be simply expressed as the following questions: Does load leveling with BESS necessarily lead to the economic benefit of a utility company? If so, how does the economic benefit change as the BESS size increases? If not, what is the reason for the results contrary to the previous studies?

Answering these questions fills a research gap as the insightfulness of previous studies is limited by the fact that electricity prices are always regarded as given. However, if the number and/or size of BESS is large, then the leveled loads with BESS will result in a change in electricity prices, which, in turn, modifies the charging/discharging operation of BESS. Accounting for this reality induces an iterative process that is challenging to resolve. If the electricity price follows a typical parabolic cost function, the iterative process may be resolved by mathematically finding a converging point. However, in a competitive electricity market, the strategic decisions of participants may distort the electricity price pattern. Therefore, the change in electricity prices due to the leveled loads with BESS should be estimated as a basis for performing any economic analysis. Consequently, in this study, we derive the answers to the research questions by conducting an economic analysis of load leveling from the perspective of a utility company, while considering changes in electricity prices, particularly thus induced by large BESS in an electricity market environment. To illustrate the results, we consider the situation of Korea's electricity market and use the corresponding data as a case study.

The contributions of this paper are (1) the change in electricity prices due to the leveled loads are explicitly considered in the economic analysis as the price function is defined with respect to loads; (2) in defining the price function, a data-driven regression model is used and estimation accuracy is achieved by data grouping; (3) it is shown as a main claim that it may be blind faith for load leveling with BESS whereby a large cumulative capacity of BESS is always economically beneficial.

The remainder of the paper is organized as follows. To provide important background information, Section 2 summarizes the structure of the Korean electricity market. Section 3 describes the optimization formulation to perform load leveling and a method to estimate electricity prices with respect to the load. Section 4 presents the necessary data for the analysis. Section 5 presents the results of the case study, followed by a discussion in Section 6. Finally, concluding remarks are provided in Section 7.

2. Korean Electricity Market

The electric industry in Korea was re-structured in 2001, whereby the vertically integrated utility company, the Korea Electric Power Corporation (KEPCO), was divided into an independent system operator, the Korea Power Exchange (KPX); some generation companies (GENCOs), subsidiaries of KEPCO; and the KEPCO itself playing correspondingly a reduced role. The Korean electricity market was also defined as a result of the restructuring and is now run by the KPX.

The Korean electricity market is a cost-based pool. The GENCOs submit the amount of electricity generation capacity and the cost function to the KPX. Then, the KPX determines the hourly generation of each GENCO based on forecasted demand. The methods applied to this end, such as economic dispatch or optimal power flow, must satisfy various conditions, such as balance of supply and demand, reserve requirements, and maximum power flows of transmission lines. The KPX also sets the hourly system marginal price (SMP). The SMP is usually set as the marginal generation cost of the most expensive generator, or the marginal generator, and is used in the settlement process. The KEPCO buys electricity through the market and pays the GENCOs according to the SMP. The KEPCO recovers the cost of purchasing electricity by invoicing customers based on the billing rates approved by the government agency.

3. Methodology

3.1. Load Leveling

The variation in original loads is mitigated or leveled by BESS charging/discharging operations. The load leveling performance can be evaluated based on how close the leveled loads are to a reference value. The mean-squared error (MSE) is selected in this study as a measure of closeness to the reference value. Thus, the objective function to determine the optimal BESS operation for load leveling is the MSE with respect to the reference value ($\overline{P}_{ref,h}$):

$$\frac{1}{H}\sum_{h=1}^{H}\left(P_{L,h} - \overline{P}_{ref,h}\right)^2 \tag{1}$$

where $P_{L,h}$ is the original load at hour h and H is the number of hours corresponding to the time horizon for load leveling. The reference, $\overline{P}_{ref,h}$, is defined for all h as the constant average of the original loads during the entire time period:

$$\overline{P}_{ref,h} = \frac{1}{H}\sum_{k=1}^{H} P_{L,k}, \quad h = 1, 2, \cdots H \tag{2}$$

Alternatively, BESS can be operated for load leveling on a daily basis. In this case, the reference value on a specific day d may be defined as:

$$\overline{P}_{ref,h} = \frac{1}{24} \sum_{k=1}^{24} P_{L,k}, \quad k \in d \qquad (3)$$

For convenience, the references in Equations (2) and (3) are denoted as the yearly reference (Yref) and the daily reference (Dref), respectively. After including BESS, $P_{L,h}$ in (1) is replaced with the leveled load, $\widetilde{P}_{L,h}$, defined as:

$$\widetilde{P}_{L,h} = P_{L,h} + P_{B,h} \qquad (4)$$

where $P_{B,h}$ is the electricity used to charge the BESS (leading to a positive value) or discharged from the BESS (resulting in a negative value) at hour h. Therefore, load leveling involves determining the value of $P_{B,h}$ that minimizes Equation (1), subject to the BESS-related constraints described below.

The first constraint is related to the maximum hourly charging and discharging power, P_B^{max}, given as:

$$-P_B^{max} \leq P_{B,h} \leq P_B^{max} \qquad (5)$$

In this study, the value of P_B^{max} is considered equal to BESS capacity. For normal operation, the state of charge (SOC) of the BESS should be bounded:

$$SOC_B^{low} \leq SOC_{B,h} \leq SOC_B^{upp} \qquad (6)$$

where $SOC_{B,h}$ is the SOC at hour h and SOC_B^{low} and SOC_B^{upp} are the lower and upper bounds of the SOC, respectively. The value of $SOC_{B,h}$ is updated every hour according to the following rule:

$$SOC_{B,h} = \begin{cases} SOC_{B,h-1} + \eta_{B,ch} P_{B,h}/P_B^{max}, & \text{if } P_{B,h} > 0 \text{ (charging)} \\ SOC_{B,h-1}, & \text{if } P_{B,h} = 0 \text{ (no operation)} \\ SOC_{B,h-1} + P_{B,h}/(\eta_{B,dch} P_B^{max}), & \text{if } P_{B,h} < 0 \text{ (discharging)} \end{cases} \qquad (7)$$

where $\eta_{B,ch}$ and $\eta_{B,dch}$ are the charging and discharging efficiencies, respectively. From Equation (7), even if $\eta_{B,ch}$ and $\eta_{B,dch}$ are equal, the update equation is asymmetric because the effects of these values are different for charging and discharging. However, the asymmetry for considering the loss of BESS makes the solution of the optimization problem using the mixed integer quadratic programming method extremely time-consuming. For example, when H is set to 168 h (one week), both $\eta_{B,ch}$ and $\eta_{B,dch}$ are set to 95%, the BESS size is 50 GWh, the computation time to find the optimal solution is 110 s. By contrast, if the loss of BESS is neglected, that is, if $\eta_{B,ch}$ and $\eta_{B,dch}$ in Equation (7) are equal to one, it can be reduced to a simpler equation:

$$SOC_{B,h} = SOC_{B,h-1} + P_{B,h}/P_B^{max} \qquad (8)$$

As a result, for the same load leveling problem over 168 h, the computation time falls significantly to just 0.2 s, which is a 99.82% decrease in computation time. The leveled loads with and without the loss of BESS are shown in Figure 1 together with the original load. Figure 1 clarifies that the solutions for the two cases are very similar in shape, and their mean absolute percentage errors (MAPEs) are as small as 0.26%. This minor difference is not the primary concern of this study. Moreover, in our analysis, H is equal to 8760 h (one year). This means that if the formulas that consider the loss of BESS are applied, the computation time would be too large because the search space expands exponentially as H increases. Consequently, Equation (8) without loss is used instead of Equation (7) for the remainder of this study. This enhances computational efficiency without causing a noteworthy loss of informativeness of results.

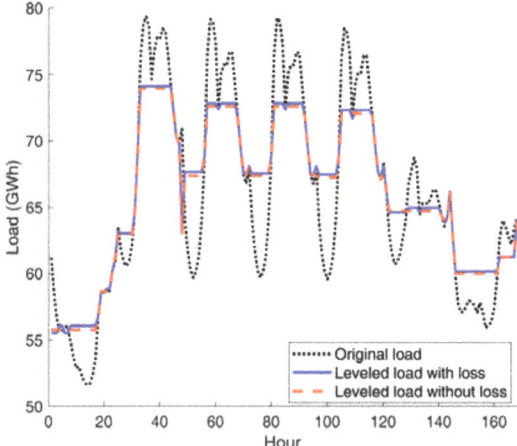

Figure 1. Leveled loads with and without the loss of battery energy storage systems (BESS) compared with original loads.

3.2. Electricity Price as a Function of Load

Electricity prices should be represented as a function of load to estimate the change in price for the leveled loads. Among the methods for identifying the relationship between input and output values, we selected the polynomial regression method [38], in which the load corresponds to the input, and the electricity price is the output. Then, the price function, $\rho_h(P_{L,h})$, can be represented as:

$$\rho_h(P_{L,h}) = \beta_0 + \beta_1 P_{L,h} + \beta_2 P_{L,h}^2 + \cdots + \beta_m P_{L,h}^m + \varepsilon_h \tag{9}$$

where β_m's are the regression coefficients and ε_h is the error term. It should be noted that the original loads, $P_{L,h}$, are used in (9) because the price function is found using the original loads. Then, the change in price after load leveling is estimated by substituting the leveled load, $\widetilde{P}_{L,h}$, into the price function in (9).

The degree of the polynomial, m, is selected according to the adjusted R^2 index, which indicates the quality of the regression model. When a model fits the data well, the adjusted R^2 index becomes closer to 1 [38]. Therefore, the degree of the polynomial is chosen as the value where the adjusted R^2 saturates at a value closer to 1 or its rate of increase becomes moderately small.

4. Case Description

The data on loads and electricity prices are taken from the Korean power system for the year 2018 [39,40]. The case study consists of three steps. The first step is to find the leveled loads using BESS based on the formulation in Section 3.1. It is performed by varying the BESS size from 1 GWh to 150 GWh by increments of 1 GWh, where the maximum size is approximately 1.7 times the peak load of 87.03 GWh. The maximum and minimum SOC are set to 90% and 10%, respectively. The two references of Yref and Dref are applied separately. The second step is to represent the electricity prices as a function of loads using a polynomial regression. In order to determine a suitable degree of polynomial, the adjusted R-squared values are calculated by varying the degree of the polynomial (which entails varying the number of terms in the polynomial regression) from one to nine. Then, the fitted price function of loads is derived as a polynomial function of the selected degree. The final step is to perform the economic analysis by calculating the total cost of leveled loads based on the results from the first and second steps.

For a more detailed economic analysis, another simulation, which is essentially the opposite of the previous scenario, is performed to examine how the BESS operation for reducing the cost affects the load leveling performance. In other words, BESS are operated with the goal of minimizing the total cost, not optimizing load leveling. Then, it can be investigated to what extent the adjusted loads are spread with respect to the reference, Yref or Dref, by calculating the MSE. In this second scenario, the first step is to again vary the BESS size from 1 GWh to 150 GWh by increments of 1 GWh, thus determining the charging/discharging schedule of BESS, which reduces the total cost for the original electricity prices. Then, the adjusted loads and the corresponding total cost are calculated. The second step is to calculate the MSE for the adjusted loads with respect to Yref and Dref.

The monthly and hourly load patterns are represented as box plots using Matlab in Figure 2a,b, respectively. When generating Figure 2a, the hourly loads over the year are grouped by month. Thus, the number of data points in each month is equal to 24 h times the number of days in the corresponding month. Then, the monthly loads in each month are sorted in ascending order and represented as a box plot, where the upper and lower sides of the blue box indicate the 75th percentile (the third quartile, Q3) and 25th percentile (the first quartile, Q1), respectively. The red line in the middle of the blue box indicates the median (50th percentile) of the sorted loads and the highest and lowest black lines indicate the maximum and minimum values, respectively. When generating Figure 2b, the hourly loads over the year are grouped by hour, such that there are 365 data for each hour group. Then, the hourly loads for each hour are sorted in ascending order and represented as a box plot. Like Figure 2a, the upper side, the red line, and the lower side of the blue box indicate the 25th percentile, 50th percentile (median), and 75th percentile of the sorted loads. However, the hourly loads are distributed quite widely, such that there are some points that are greater than Q3 + 1.5 times (Q3 − Q1) or less than Q1 − 1.5 times (Q3 − Q1). These data points are referred to as whiskers. This type of points can be interpreted as outliers in data analysis and are represented as red crosses in Figure 2b. If there are red crosses outside the highest and lowest black lines, then the black line indicates Q3 + 1.5 times (Q3 − Q1) or Q1 − 1.5 times (Q3 − Q1); otherwise, it indicates the maximum or minimum value of the sorted hourly loads. The average hourly load of the analyzed year is equal to 61.25 GWh. Figure 2 shows that summer (July and August) and winter (January, February, and December) are the peak seasons of electricity consumption over the year. In addition, during a day, two peaks occur just before and just after noon.

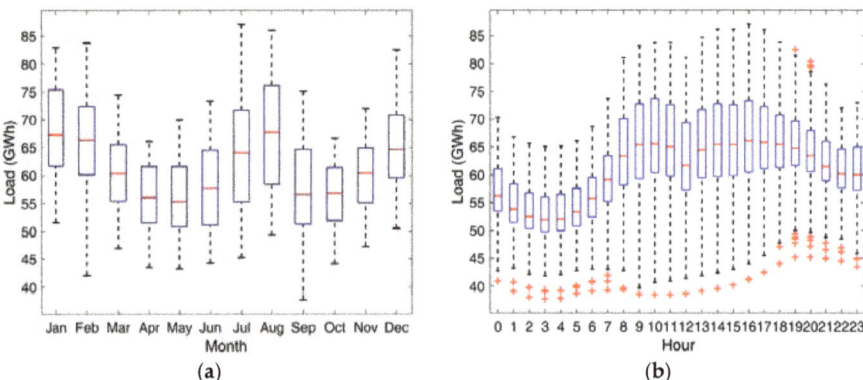

Figure 2. Box plots for original loads: (a) Monthly loads; (b) Hourly loads.

The monthly and hourly patterns of original electricity prices are represented as the box plots using Matlab in Figure 3a,b, respectively. The methods of grouping the price data for the year and the meaning of the symbols of the box plot are the same as those for the loads in Figure 2a,b. Electricity prices need to be represented as a function of load in this study, which requires a relationship to exist between them. Figure 3b shows a similar pattern to Figure 2a, although peak and off-peak hours are not as apparent as the loads. Therefore, it can be inferred that there is a positive relationship between loads and electricity prices. However, the monthly variation of electricity prices in Figure 3a does not show a clear pattern which was observed seen in Figure 2a. Therefore, we checked the correlation coefficient between loads and electricity prices. The correlation coefficient for the entire dataset is 0.47, indicating a moderately positive correlation. By contrast, Table 1 shows that the monthly correlation coefficients are very high, indicating a highly positive relationship. Consequently, the polynomial regression is performed on monthly data, so that a total of 12 price functions are found to estimate the price change in response to the leveled loads. The specific results of the polynomial regression are presented in Section 4.

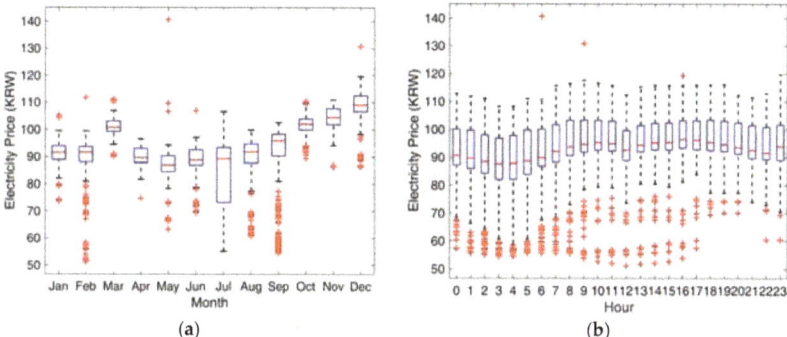

Figure 3. Box plots for original electricity prices: (a) Monthly prices; (b) Hourly prices.

Table 1. Correlation coefficients between monthly loads and electricity prices.

January	February	March	April	May	June
0.84	0.79	0.81	0.92	0.72	0.85
July	August	September	October	November	December
0.90	0.84	0.81	0.88	0.87	0.84

5. Simulation Results

5.1. Load Leveling

It is expected that the load will be leveled more and the MSE will decrease as the BESS size increases. Further, the important result to be examined is not the absolute value of the MSE, but how much the MSE is reduced when BESS size is increased. Therefore, the load leveling performance in terms of the MSE is represented in Figure 4a as the percentage ratio to the MSE of the original loads without BESS. Since two references of Y_{ref} and D_{ref} are used in this study, two graphs of the MSE ratio are shown in Figure 4a. Additionally, the original loads and the leveled loads with the BESS sizes of 50 GWh and 150 GWh for the month of May are shown in Figure 5 to identify the changes from load leveling more clearly. Another measure to examine the load leveling performance is the maximum/minimum values of the leveled loads because they will be closer to a reference, that is, the maximum will decrease and the minimum will increase, as the BESS size increases. Thus, the maximum/minimum values of the leveled loads for Y_{ref} and D_{ref} are given in Figure 4b.

Figure 4a shows that the marginal improvement in the load leveling performance by the unit increase in BESS size gradually decreases as BESS size increases. This is because the load duration of very high or very low loads is relatively short, so that a relatively small BESS can significantly improve load leveling. However, as BESS size increases, the duration for which loads need to be lowered or raised with respect to the reference value increases. Therefore, the degree of reduction in MSE decreases. More specifically, the decrease in the degree of reduction is severe, especially for Y_{ref}, because the load leveling is performed for the same constant reference over the entire time period. Figure 4a clarifies this relationship by showing the slow decrease in the MSE for Y_{ref}, which makes a significant difference for a large BESS. Specifically, the MSE for Dref is reduced to less than 10% for a BESS size greater than 70 GWh. By contrast, the MSE for Yref is still equal to approximately 48%, even for a BESS size of 150 GWh.

On the other hand, as Figure 4b shows, the maximum and minimum values of the leveled loads are almost the same for the two references. Specifically, the maximum values of the leveled loads for the two references are equal to each other in the beginning (hence why the lines in the graph almost exactly coincide), but a difference in the minimum values becomes evident from the BESS size of 30 GWh. The results can be clarified with Figure 5. Regardless of whether the load leveling is performed during a year or a day, the goal is to lower the high loads and raise the low loads. Thus, the overall shapes of Figure 5 for the two reference values are similar. However, the highs and lows improve more for Yref than for Dref. This is because the number of days with high loads, such as summer and winter weekdays, is greater than the number of days with low loads, such as holidays and spring/autumn weekends. Therefore, lowering the maximum total loads for Yref is equivalent to lowering the maximum of a day of high loads for Dref. However, after increasing the minimum of the holidays and weekends of low loads, the minimum of the leveled days of low loads remains unchanged. Consequently, as seen in Figure 4b, the minimum of the leveled load for Dref hardly changes after some point, which corresponds to a BESS size of 30 GWh in this case study.

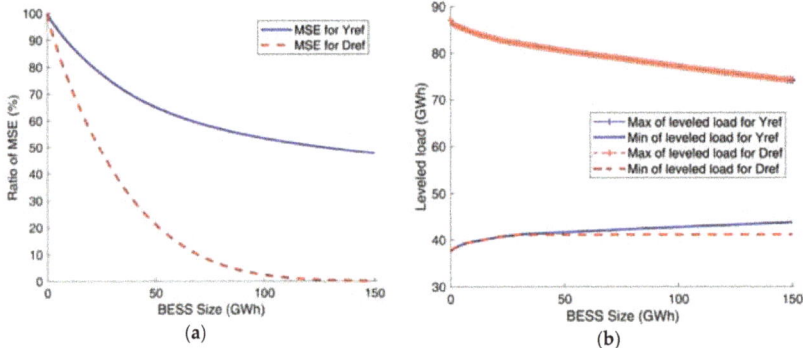

Figure 4. Mean-squared error (MSE) and maximum/minimum values of the leveled loads with respect to BESS size: (**a**) MSE for Yref and Dref; (**b**) Maximum and minimum values for Yref and Dref.

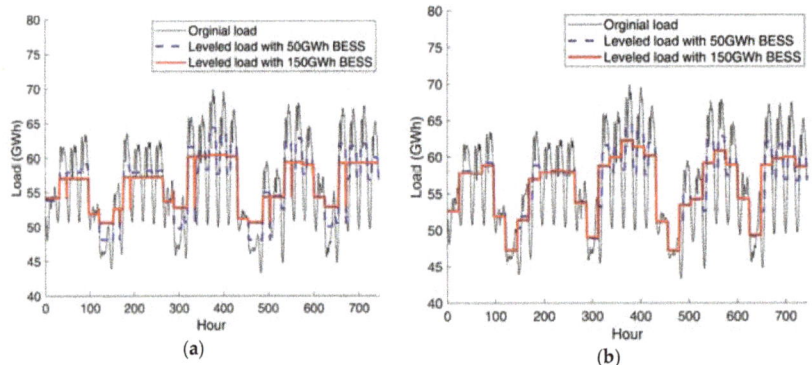

Figure 5. Original and leveled loads with BESS of 50 GWh and 150 GWh in May: (**a**) Leveled loads for Yref; (**b**) Leveled loads for Dref.

5.2. Representation of Electricity Prices

In Figure 6, electricity prices are represented as a scatter plot with respect to the load. Although the electricity price appears to increase as the load increases, a relationship between them is difficult to identify. Therefore, considering the monthly correlation coefficients in Table 1, we divide the data by month and represent them as separate scatter plots in Figure 7. Contrary to Figure 6, the variation of data points is small, so a relationship between electricity prices and loads becomes evident.

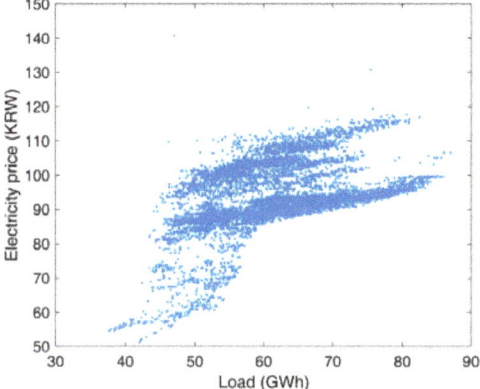

Figure 6. Electricity prices with respect to the loads over a year.

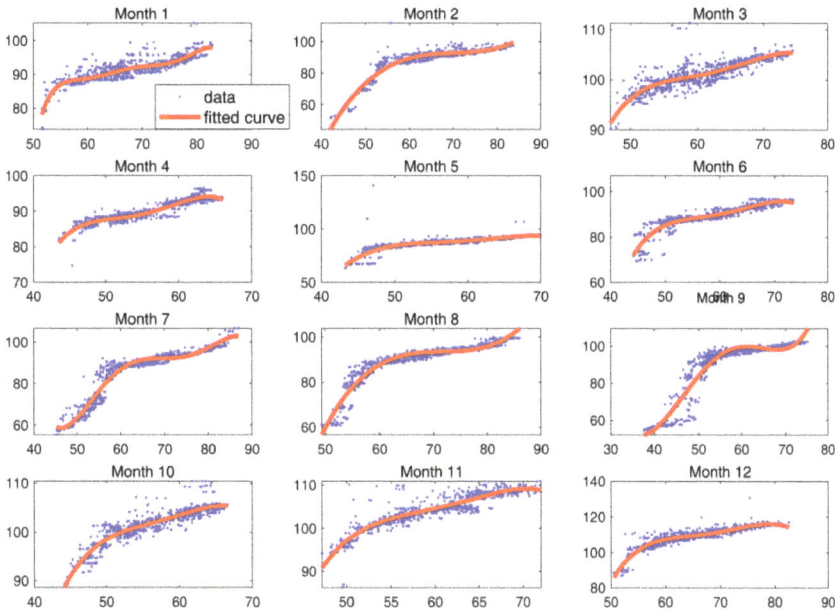

Figure 7. Electricity prices with respect to loads (blue dots) in each month and the fitted prices (solid red lines) using the polynomial regression.

The results of the adjusted R-squared values are shown in Figure 8. By checking whether the adjusted R-squared value is close enough to 1, the degrees of the polynomial of each month, from January to December, are selected as 6, 3, 4, 4, 4, 4, 5, 3, 4, 4, 4, and 4, respectively. The fitted functions are represented as the solid red lines on the original data points in Figure 7. The fitted graph of each month in Figure 7 represents the original electricity prices well. Specifically, the MAPE of the fitted electricity prices during a year is equal to 1.78%.

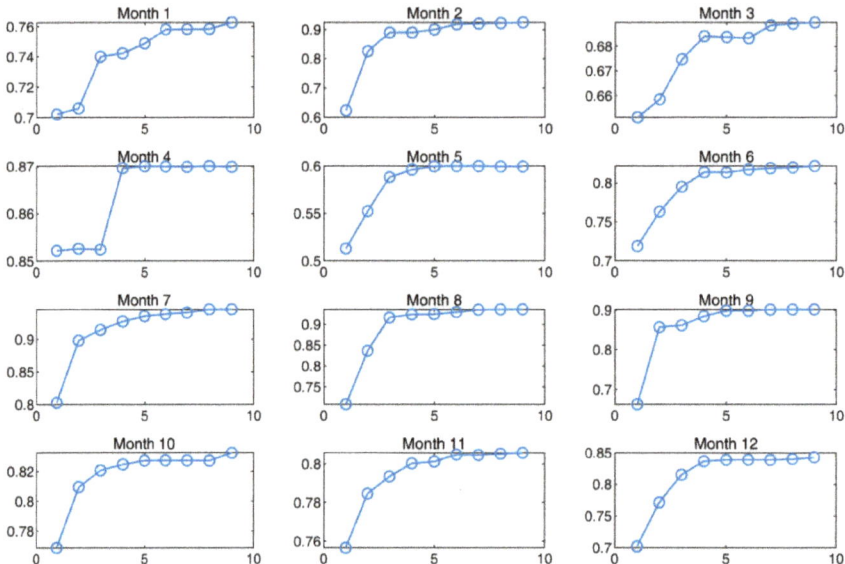

Figure 8. Adjusted R-squared values of the polynomial regression with a degree of polynomial from one to nine.

5.3. Economic Analysis

First, the accuracy of the fitted electricity price function is examined in terms of the total cost. The total cost for the original prices and loads is calculated as 50,776 billion KRW and the total cost for the fitted prices and original loads is also equal to 50,776 billion KRW. This is evidence of the suitability of the fitted price function. The estimated costs after load leveling with the BESS are shown in Figure 9 as a percentage of the total cost of 50,776 billion KRW for the original prices and loads without BESS. Figure 9 shows that, regardless of whether the load leveling is performed for Yref or Dref, the total cost monotonically increases as BESS size increases. Based on the expectation that an economic benefit should accrue from load leveling due to the reduction in peaks, this result seems counterintuitive at first. However, closer inspection of Figure 7 indicates that this should nevertheless be the natural result because electricity prices increase fast in the range of small loads, but they increase slowly in the range of high loads. As a result, the increased cost due to upward leveling of small loads is greater than the decreased cost from downward leveling of high loads. Consequently, the total cost increases after load leveling with BESS. In addition, Figure 7 shows that the total cost for Yref is greater than that for Dref. This implies that more effort is required to flatten the loads over a longer time period, resulting in a greater cost increase.

The results suggest that we need to rethink the purpose of load leveling. If the price function is a quadratic one based on the cost of the physical operation of generators [41], the total cost will monotonically decrease as BESS size increases, although the extent of the decrease will gradually become smaller. Then, it should be a natural decision to use BESS for load leveling, and the problem is essentially reduced to deciding on a suitable BESS size by comparing the cost decrease from the leveled loads against the cost increase from the BESS. However, as Figure 9 indicates, the costs which are physically incurred by generators is not the only factor to be considered in deciding on electricity prices or bidding prices in a market environment. For example, under some bidding strategies, bid prices for low loads could be set much higher than the actual generation costs and those for high loads could be set at a level comparable to the actual generation cost. In this case, the electricity prices shown in Figure 7 would result. Consequently, without other benefits, such as improved reliability and stability due to the leveled load, one cannot simply assume that load leveling is a worthwhile grid application of BESS from an economic standpoint, particularly in an electricity market environment.

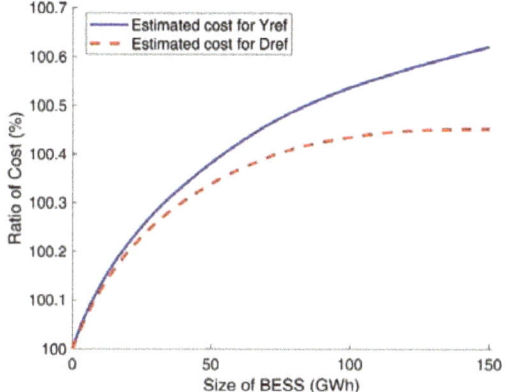

Figure 9. Estimated cost of leveled loads with respect to BESS size.

5.4. BESS Operation for Reducing the Cost

The results of the second scenario, the cost minimization scenario, are shown in Figure 10. Like Figure 9, the total cost and the MSE are represented in Figure 10 as the percentage of the corresponding results for the original prices and loads without BESS. Figure 10 shows that the total cost decreases linearly as the BESS size increases because BESS are operated to minimize the total cost. However, the MSEs for Yref and Dref increase rapidly to tens of times the MSE without BESS, which is completely against the purpose of load leveling. This is because, in the process of minimizing the cost, the loads greater than the reference value are increased and the loads less than the reference value are decreased. This means that the loads can be changed to deviate further from the reference value to reduce the cost.

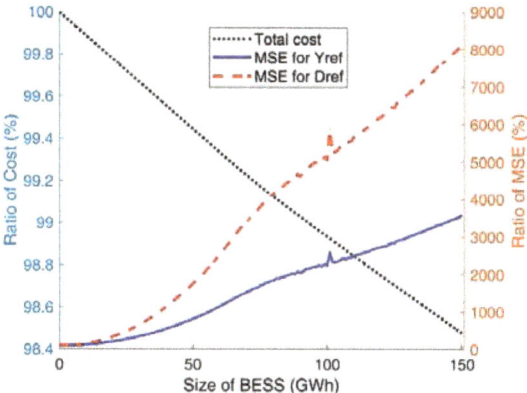

Figure 10. Total cost from optimized BESS operation to reduce the cost and MSEs for Yref and Dref of the resulting loads.

6. Discussion

The results of load leveling in Section 5.1 indicate that loads can be flattened more as BESS size increases. However, it should be noted that the extent of load leveling performance is gradually reduced, while the costs of BESS remain high. Therefore, if the total cost of electricity consumption decreases due to the leveled loads, it may be necessary to compare the marginal reduction in the total cost reduction with the marginal increase in the investment for the unit increase in BESS size. Then, the optimal BESS size can be determined. However, as presented in Section 5.3, the total cost rather monotonically increases as the BESS size increases and the load is correspondingly flattened more. This suggests that the load leveling with BESS may not necessarily lead to economic benefits for a utility company, particularly in a competitive electricity market environment. Subsequently, it becomes meaningless to discuss how much the economic benefit is as the BESS size increases and which BESS size is optimal for the load leveling application in terms of economic benefit. What is the reason for these counterintuitive results? It can be inferred from Figure 7 that the complex strategic decisions of the participants in an electricity market may result in a complex pattern of electricity prices, whereby prices increase quickly in the range of small loads but increase slowly in the range of high loads. In this case, it may be simply incorrect to state that the utility's investment in BESS can be recovered after some years through the reduced purchasing cost of electricity.

This conflict between the economic benefit and the load leveling application with BESS can be verified more clearly by reviewing the results of the second scenario. In order to reduce the total cost, BESS are charged during the periods of low electricity price and discharged during high price periods. However, this means that operating BESS with the goal of reducing electricity costs actually widened the variation of the resulting loads. Consequently, in a certain electricity market environment, the economic operation of BESS to recover investment costs will increase the fluctuation of net loads. As a result, the side costs of maintaining the reliability and stability of electricity networks may undermine the utility company's profit.

The results in this study lead to some guidelines for designing a policy related to the BESS for load leveling. First, it should not simply be assumed that BESS with a high cumulative capacity is always economically beneficial even if the investment cost in BESS is low enough. Second, load leveling with the BESS may rather increase the total cost of electricity provision, particularly in a competitive electricity market environment. In this case, the load leveling needs to be linked to other purposes than economic benefit to justify the validity of using the BESS. Third, when a large number of distributed BESS are integrated into the power system, the pricing mechanism of electricity should be elaborately designed to prevent the synchronized operation of the BESS. This is necessary to ensure systems remain sufficiently profitable and that the power system is reliable.

7. Conclusions

BESS, alongside renewable energy sources, are expected to become a major component of future power systems. However, an analysis of their economic feasibility is essential because, at present, the costs of investing in BESS are significant. Among the various grid applications of BESS, this study focused on load leveling and analyzed its economic benefits in an electricity market environment. For economic analysis, load leveling is formulated as a constrained optimization problem, which is then solved using mixed integer quadratic programming. Additionally, the electricity prices are represented as a function of the load using the polynomial regression method to estimate the adjusted prices for the leveled loads.

The results of the economic analysis using Korean data show that the degree of reduction in the MSEs of the leveled loads decreases as BESS size increases. The problem that MSE reduction decreases with the increase in BESS size is more severe for the yearly reference because load leveling is performed for the same constant reference over a longer time period. Further, the results show that, contrary to the simple expectation that load leveling will result in decreased cost of electricity generation, the total cost monotonically increases as the BESS size increases and accordingly, the loads are flattened more. By contrast, if BESS are operated to minimize the cost, the variation of net loads severely increases, which is the opposite to what load leveling is intended to achieve.

The results suggest some guidelines for designing policy related to BESS for load leveling. First, it should not be assumed that a large cumulative capacity of BESS is always economically beneficial if its price is low enough. Second, load leveling with BESS may actually increase the total cost of electricity provision, particularly in a competitive electricity market environment. In this case, the justification for pursuing load leveling needs to be based on some purpose other than economic benefit to ensure the validity of using BESS. Finally, when a large number of distributed BESS are integrated into the power system, the electricity pricing scheme should be elaborately designed to prevent the synchronized operation of BESS and thereby, ensure the power system remains reliable.

Meaningful further research could consider the following topics: First, it is assumed that GENCOs' bids do not change for utilities' application of load leveling with BESS. That is, the price function itself may change with leveled loads, which results in another iterative process. Thus, defining this iterative mechanism becomes a research issue, which will be very difficult to be addressed because it involves strategic, not just technical, decisions on behalf of GENCOs. Second, if the leveled loads improve reliability and stability of power systems, a different kind of economic analysis can be done by translating the improved reliability into monetary value. Third, it is assumed in this study that BESS are operated only for load leveling. Thus, it may be necessary to consider that BESS are utilized partly for load leveling and partly for another purpose, and to determine an optimal proportion between the applications of BESS.

Author Contributions: H.T.K. developed the optimization model, set up the simulation environment, and performed the simulations. Y.G.J. designed the study, performed the analysis, thoroughly revised the paper and checked the overall logic of the work. Y.T.Y. provided insightful comments on the modeling and analysis.

Funding: This research was supported by Korea Electric Power Corporation (Grant number: R18XA03) and the 2019 scientific promotion program funded by Jeju National University.

Conflicts of Interest: The authors declare no conflict of interest.

Abbreviations

Acronyms

RES	Renewable energy source
PV	Photovoltaic
ESS	Energy storage systems
BESS	Battery energy storage systems
EV	Electric vehicle
NPV	Net present value
KEPCO	Korea Electric Power Corporation
KPX	Korea Power Exchange
GENCO	Generation company
SMP	System marginal price
MSE	Mean-squared error
SOC	State of charge
MAPE	Mean absolute percentage error

Indices

h	Index of hour
d	Index of day

Parameters

$x\overline{P}_{ref,h}$	Reference for load leveling
$P_{L,h}$	Original load at hour h
P_B^{max}	BESS size
SOC_B^{low}	Lower bound of the SOC
SOC_B^{upp}	Upper bounds of the SOC
$\eta_{B,ch}$	Charging efficiency of BESS
$\eta_{B,dch}$	Discharging efficiency of BESS

H	Number of hours
m	Degree of polynomial in regression
Variables	
$\widetilde{P}_{L,h}$	Leveled load at hour h
$P_{B,h}$	Electricity charged (positive) to or discharged (negative) from BESS at hour h
$SOC_{B,h}$	SOC of BESS at hour h
ρ_h	Electricity price at hour h
β_m	Regression coefficients
ε_h	Error term in regression

References

1. Global Wind Energy Council. Global Wind Report-Annual Market Update 2017. Available online: http://gwec.net/publications/global-wind-report-2/ (accessed on 6 July 2018).
2. International Energy Agency. A Snapshot of Global PV (1992–2017). Available online: http://www.iea-pvps.org/?id=266 (accessed on 6 July 2018).
3. Xie, L.; Carvalho, P.M.S.; Ferreira, L.A.F.M.; Liu, J.; Krigh, B.H.; Popli, N.; Ilic, M.D. Wind Integration in Power Systems: Operational Challenges and Possible Solutions. *Proc. IEEE* **2011**, *99*, 214–232. [CrossRef]
4. Castillo, A.; Gayme, D.F. Grid-Scale Energy Storage Applications in Renewable Energy Integration: A Survey. *Energy Convers. Manag.* **2014**, *87*, 885–894. [CrossRef]
5. Luo, X.; Wang, J.; Dooner, M. Overview of Current Development in Electrical Energy Storage Technologies and the Application Potential in Power System Operation. *Appl. Energy* **2015**, *137*, 511–536. [CrossRef]
6. IRENA. Electricity Storage and Renewables: Costs and Markets to 2030. International Renewable Energy Agency (IRENA), 2017. Available online: https://www.irena.org/publications/2017/Oct/Electricity-storage-and-renewables-costs-and-markets (accessed on 25 February 2019).
7. U.S. Energy Information Administration. Battery Storage Market Trends. U.S. Energy Information Administration, 2018. Available online: https://www.eia.gov/analysis/studies/electricity/batterystorage/ (accessed on 25 February 2019).
8. Coignard, J.; Saxena, S.; Greenblatt, J.; Wang, D. Clean Vehicles as an Enabler for a Clean Electricity Grid. *Environ. Res. Lett.* **2018**, *13*, 054031. [CrossRef]
9. Bloomberg NEF. Energy Storage is a $620 Billion Investment Opportunity to 2040. 2018. Available online: https://about.bnef.com/blog/energy-storage-620-billion-investment-opportunity-2040/ (accessed on 19 February 2019).
10. Xia, S.; Chan, K.W.; Luo, X.; Bu, S.; Ding, Z.; Zhou, B. Optimal Sizing of Energy Storage System and Its Cost-Benefit Analysis for Power Grid Planning with Intermittent Wind Generation. *Renew. Energy* **2018**, *122*, 472–486. [CrossRef]
11. Korpaas, M.; Holen, A.T.; Hildrum, R. Operation and Sizing of Energy Storage for Wind Power Plants in a Market System. *Int. J. Electr. Power Energy Syst.* **2003**, *25*, 599–606. [CrossRef]
12. Ru, Y.; Kleissl, J.; Martinez, S. Storage Size Determination for Grid-Connected Photovoltaic Systems. *IEEE Trans. Sustain. Energy* **2013**, *4*, 68–81. [CrossRef]
13. Harsha, P.; Dahleh, M. Optimal Management and Sizing of Energy Storage Under Dynamic Pricing for the Efficient Integration of Renewable Energy. *IEEE Trans. Power Syst.* **2015**, *30*, 1164–1181. [CrossRef]
14. Khalilpour, R.; Vassallo, A. Planning and Operation Scheduling of PV-Battery Systems: A Novel Methodology. *Renew. Sustain. Energy Rev.* **2016**, *53*, 194–208. [CrossRef]
15. Talent, O.; Du, H. Optimal Sizing and Energy Scheduling of Photovoltaic-Battery Systems under Different Tariff Structures. *Renew. Energy* **2018**, *129*, 513–526. [CrossRef]
16. Telaretti, E.; Ippolito, M.; Dusonchet, L. A Simple Operating Strategy of Small-Scale Battery Energy Storages for Energy Arbitrage under Dynamic Pricing Tariffs. *Energies* **2016**, *9*, 12. [CrossRef]
17. Hemmati, R. Technical and Economic Analysis of Home Energy Management System Incorporating Small-Scale Wind Turbine and Battery Energy Storage System. *J. Clean. Prod.* **2017**, *159*, 106–118. [CrossRef]

18. Lujano-Rojas, J.M.; Dufo-López, R.; Bernal-Agustín, J.L.; Catalão, J.P.S. Optimizing Daily Operation of Battery Energy Storage Systems Under Real-Time Pricing Schemes. *IEEE Trans. Smart Grid* **2017**, *8*, 316–330. [CrossRef]
19. Faqiry, M.N.; Edmonds, L.; Zhang, H.; Khodaei, A.; Wu, H. Transactive-Market-Based Operation of Distributed Electrical Energy Storage with Grid Constraints. *Energies* **2017**, *10*, 1891. [CrossRef]
20. Fleer, J.; Zurmühlen, S.; Meyer, J.; Badeda, J.; Stenzel, P.; Hake, J.-F.; Sauer, D.W. Price Development and Bidding Strategies for Battery Energy Storage Systems on the Primary Control Reserve Market. In Proceedings of the 11th International Renewable Energy Storage Conference, IRES 2017, Düsseldorf, Germany, 14–16 March 2017.
21. Xiang, Y.; Liu, J.; Li, R.; Li, F.; Gu, C.; Tang, S. Economic Planning of Electric Vehicle Charging Stations Considering Traffic Constraints and Load Profile Templates. *Appl. Energy* **2016**, *178*, 647–659. [CrossRef]
22. Baik, S.H.; Jin, Y.G.; Yoon, Y.T. Determining Equipment Capacity of Electric Vehicle Charging Station Operator for Profit Maximization. *Energies* **2018**, *11*, 2301. [CrossRef]
23. Luo, L.; Gu, W.; Zhou, S.; Huang, H.; Gao, S.; Han, J.; Wu, Z.; Dou, X. Optimal Planning of Electric Vehicle Charging Stations Comprising Multi-Types of Charging Facilities. *Appl. Energy* **2018**, *226*, 1087–1099. [CrossRef]
24. Nikolovski, S.; Baghaee, H.R.; Mlakic, D. ANFIS-Based Peak Power Shaving/Curtailment in Microgrids Including PV Units and BESSs. *Energies* **2018**, *11*, 2953. [CrossRef]
25. Kaviani, A.K.; Baghaee, H.R.; Riahy, G.H. Optimal Sizing of a Stand-Alone Wind/Photovoltaic Generation Unit Using Particle Swarm Optimization. *Simulation* **2009**, *85*, 89–99. [CrossRef]
26. Baghaee, H.R.; Mirsalim, M.; Gharehpetian, G.B.; Talebi, H.A. Reliability/Cost-Based Multi-Objective Pareto Optimal Design of Stand-Alone Wind/PV/FC Generation Microgrid System. *Energy* **2016**, *115*, 1022–1041. [CrossRef]
27. Palizban, O.; Kauhaniemi, K. Energy Storage Systems in Modern Grids—Matrix of Technologies and Applications. *J. Energy Storage* **2016**, *6*, 248–259. [CrossRef]
28. Lo, C.H.; Anderson, M.D. Economic Dispatch and Optimal Sizing of Battery Energy Storage Systems in Utility Load-Leveling Operations. *IEEE Trans. Energy Convers.* **1999**, *12*, 824–829. [CrossRef]
29. Kerestes, R.J.; Reed, G.F.; Sparacino, A.R. Economic Analysis of Grid Level Energy Storage for the Application of Load Leveling. In Proceedings of the 2012 IEEE Power and Energy Society General Meeting, San Diego, CA, USA, 22–26 July 2012; pp. 1–9.
30. Hemmati, R.; Saboori, H. Short-Term Bulk Energy Storage System Scheduling for Load Leveling in Unit Commitment: Modeling, Optimization, and Sensitivity Analysis. *J. Adv. Res.* **2016**, *7*, 360–372. [CrossRef] [PubMed]
31. Dicorato, M.; Forte, G.; Pisani, M.; Trovato, M. Planning and Operating Combined Wind-Storage System in Electricity Market. *IEEE Trans. Sustain. Energy* **2012**, *3*, 209–217. [CrossRef]
32. Cho, S.M.; Shin, H.S.; Kim, J.C. Modeling of Battery Energy Storage System at Substation for Load Leveling and Its Economic Evaluation. *Trans. Korean Instit. Electric. Eng.* **2012**, *61*, 950–956. [CrossRef]
33. Dupont, G.; Baltus, P. Dimensioning and Grid Integration of Mega Battery Energy Storage System for System Load Leveling. In Proceedings of the 2009 IEEE Bucharest PowerTech, Bucharest, Romania, 28 June–2 July 2009.
34. Zheng, M.; Meinrenken, C.; Lackner, K.S. Smart Households: Dispatch Strategies and Economic Analysis of Distributed Energy Storage for Residential Peak Shaving. *Appl. Energy* **2015**, *147*, 246–257. [CrossRef]
35. Telaretti, E.; Dusonchet, L. Battery Storage Systems for Peak Load Shaving Applications: Part 1: Operating Strategy and Modification of the Power Diagram. In Proceedings of the 2016 IEEE 16th International Conference on Environment and Electrical Engineering (EEEIC), Florence, Italy, 7–10 June 2016.
36. Graditi, G.; Ippolito, M.G.; Telaretti, E.; Zizzo, G. Technical and Economical Assessment of Distributed Electrochemical Storages for Load Shifting Applications: An Italian Case Study. *Renew. Sustain. Energy Rev.* **2016**, *57*, 515–523. [CrossRef]
37. Baghaee, H.R.; Mirsalim, M.; Gharehpetian, G.B. Multi-Objective Optimal Power Management and Sizing of a Reliable Wind/PV Microgrid with Hydrogen Energy Storage Using MOPSO. *J. Intell. Fuzzy Syst.* **2017**, *32*, 1753–1773. [CrossRef]
38. James, G.; Witten, D.; Hastie, T.; Tibshirani, R. *An Introduction to Statistical Learning with Applications in R*, 7th ed.; Springer: New York, NY, USA, 2013.

39. Hourly Load Forecast of Korea Power eXchange. Available online: http://www.kpx.or.kr/www/contents.do?key=223 (accessed on 18 January 2019).
40. System Marginal Price of Korea Power eXchange. Available online: http://www.kpx.or.kr/www/contents.do?key=225 (accessed on 18 January 2019).
41. Bergen, A.R.; Vital, V. *Power System Analysis*, 2nd ed.; Prentice-Hall: Upper Saddle River, NJ, USA, 2000.

© 2019 by the authors. Licensee MDPI, Basel, Switzerland. This article is an open access article distributed under the terms and conditions of the Creative Commons Attribution (CC BY) license (http://creativecommons.org/licenses/by/4.0/).

Article

Intelligent Control of Converter for Electric Vehicles Charging Station

Mayank Jha [1], Frede Blaabjerg [2,*], Mohammed Ali Khan [1], Varaha Satya Bharath Kurukuru [1] and Ahteshamul Haque [1]

1. Department of Electrical Engineering, Jamia Millia Islamia, New Delhi 110025, India; mayank.jha.595@gmail.com (M.J.); mak1791@gmail.com (M.A.K.); kvsb272@gmail.com (V.S.B.K.); ahaque@jmi.ac.in (A.H.)
2. Department of Energy Technology, Aalborg University, Aalborg 9220, Denmark
* Correspondence: fbl@et.aau.dk

Received: 28 May 2019; Accepted: 14 June 2019; Published: 18 June 2019

Abstract: Electric vehicles (EVs) are envisaged to be the future transportation medium, and demonstrate energy efficiency levels much higher than conventional gasoline or diesel-based vehicles. However, the sustainability of EVs is only justified if the electricity used to charge these EVs is availed from a sustainable source of energy and not from any fossil fuel or carbon generating source. In this paper, the challenges of the EV charging stations are discussed while highlighting the growing use of distributed generators in the modern electrical grid system. The benefits of the adoption of photovoltaic (PV) sources along with battery storage devices are studied. A multiport converter is proposed for integrating the PV, charging docks, and energy storage device (ESD) with the grid system. In order to control the bidirectional flow between the generating sources and the loads, an intelligent energy management system is proposed by adapting particle swarm optimization for efficient switching between the sources. The proposed system is simulated using MATLAB/Simulink environment, and the results depicted fast switching between the sources and less switching time without obstructing the fast charging to the EVs.

Keywords: Grid Connected Photovoltaic Systems (GCPVS); Energy Storage Device (ESD); Electric Vehicle (EV); Multiport Converter (MPC); Intelligent Energy Management System (iEMS)

1. Introduction

The current environmental challenges of reducing greenhouse gases and the potential shortage of fossil fuels motivate widespread research on electric vehicle (EV) systems [1]. However, the research on EVs is highly impacted by the consumer disposition for switching to EVs as an alternative for conventional internal combustion engine vehicles. This willingness is the main factor in forecasting future demand for EVs. In Reference [2], the authors depicted that the charging time is one of the main challenges that the EV industry is facing. Generally, the EV charging levels are classified according to their power charging rates [3]. Overnight charging takes place in level I, as the EVs are plugged to a convenient power outlet (120 V) for slow charging (1.5–2.5 kW) over long hours. The main concern of level-I is the long charging time, which renders this charging level unsuitable for long driving cycles when more than one charging operation is needed. Moreover, from the electrical grid operation point of view, the long charging hours at night overloads the distribution transformers as they are not allowed to rest in a grid system with a high number of connected EVs [4]. Level-II charging requires a 240 V outlet; thus, it is characteristically used as the prime charging means for public and private facilities. This charging level is capable of supplying power in the range of 4–6.6 kW over a period of 3–6 h to restock the depleted EV batteries. The time required is still the main drawback in this charging level. Additionally, voltage sags and high-power losses in an electrical grid system with a high penetration of level II charging are some of the challenges that are facing its widespread. Control and coordination in level II would reduce the negative impacts of level-II charging [5]; however, this requires an extensive communication system to be adopted.

In general, both levels-I and II require single-phase power sources with onboard vehicle chargers. On the contrary, three-phase power systems are used with off-board chargers for level III fast charging rates (50–75 kW). The use of fast charging stations significantly reduces the EV charging time for a complete charging cycle. Additionally, widespread deployment of fast EV charging stations across the urban and the residential areas would eliminate the EV range anxiety concern [6,7]. However, the high-power charging rates are essential over a short interval of time for level-III charging impose a very high demand on the utility grid [8,9]. The current grid infrastructure is not capable of supporting the desired high charging rates of level-III. Thus, accomplishing fast charging rates while solely depending on the electrical grid does require not only the improvement of the charging system, but also the improvement of the electrical grid capacity. Additionally, drawing large amounts of current from the electrical grid will increase the utility charges especially at the peak hours and consequently will increase the system cost. The impact of an EV charging station load on electric grid systems is thoroughly discussed in Reference [10].

A possible solutions to these challenges could be the installation of a distributed generator (DG) near the fast charging station site, as it generates electrical supply that is projected towards on-site use [11]. A few decades ago, a number of developments began to change the basics of operation of the electrical grid industry leading to the rise of DGs. The ambitious targets of the higher deployment rates of the DGs into the electrical pool is achievable, due to advancement of technologies, and the enhancements, in the fields of power electronics and smart grids. Additionally, new regulations and policies are continuously issued favoring the distributed generation and the net metering concepts. However, the type of energy used in fueling the deployed DG sources on the demand side is a decisive factor in the economic viability of the DGs concept in today's electrical distribution market.

Regarding the distributed generation, renewable energy sources (RES) have a distinct advantage in their ability to be deployed in residential and urban areas, due to their environmentally friendly operation and minimal maintenance requirements. Consequently, photovoltaics (PV) are considered as an effective solution, due to its sustainability and ubiquity. The major advantage with PV systems is, they are effectively utilized for different power generation levels starting from low-power domestic applications to mega power PV based power plants. The grid-connected PV installations are utilized with deeper penetrating levels compared to the standalone PV installations. This is due to the continuous reliance on the electric grid as a stable source/load that can compensate for the PV power fluctuations. In Reference [12], research by George et al. simulated a solar energy-based charging station by considering the decongestive knots in the capacity of network distribution systems.

In the proposed system, PV operates as grid-tied DG source. Solar energy is a preferable DG source in EV charging applications for two reasons:

1. The PV panels are more effective than other renewables in populated and residential areas, due to their noise-free operation and low maintenance requirements.
2. PV panels generate most of their energy during the highly priced grid tariff hours of the electrical grid. Thus, the EV charging stations can offset the high-cost electricity with solar energy during peak hours.

The PV power production is highly impacted by the ambient temperature and solar insolation levels, which causes discontinuity of operation. Consequently, connecting the PV panels directly to the load without any subsidiary systems leads to a negative impact on the performance of the connected electrical loads. One of the applicable solutions for the aforementioned challenge is the design of a hybrid RESs system using various RESs, which can relatively offset a portion of the local fluctuations in every generation unit. However, this requires an appropriate selection of the most suitable generation technologies, as well as a proper sizing of the (RES) [13]. The combination of PV sources with wind energy is explored in Reference [14]. However, such a system requires either the insertion of storage devices or a connection point to the electrical grid in order to support the necessary loads continuously. The authors in Reference [15] presented the use of a PV source with fuel cells to meet the requirements of a residential load. In doing so, a reserve capacity had to be maintained at the PV source to supply the load changes as the fuel cells technology instills a slow dynamic response. This leads to a deviation between the PV maximum power point (MPP) and the system operating point. A hybrid system, composed of PV/fuel-cells/ultra-capacitors, is used in Reference [16] to accommodate these challenges. In this study, the ultracapacitors are selected, due to their fast-dynamic response, which leads to their ability to mitigate the rapid fluctuations of the PV power while continuously tracking the PV MPP. However, this configuration fails to meet the load demands during extended hours of low insolation level conditions (e.g., night hours or shady days).

Under such scenarios, storage devices can play a key role in the integrated RES. The literature contains different examples of RES combined with energy storage systems. Various energy storage technologies are studied in the literature to be connected with the wind generators, hydrogen cells, and ultracapacitors [17–19] in order to provide steady output power to the electrical loads.

The authors in Reference [20] introduced the use of distributed batteries in a grid-tied PV power plant to improve energy production. Achieving a constant power production from the PV power plant is the main defined objective of inserting an energy storage system, as shown in Reference [21]. In Reference [22], a storage battery is added to the grid-connected PV system to reduce the PV power fluctuations in a defined range within a particular period of time while maximizing the revenue. This is motivated by meeting the utilities regulations and restrictions on the PV power injected to the grid.

Therefore, from the literature, the advantages of integrating battery storage devices with PV power in order to attain a stable power supply at a minimal operating cost is observed. In this research, energy storage batteries are coupled with PV panels in a grid-tied system. The proposed hybrid system is designed to provide means of fast charging for EVs. In this paper, energy storage devices, such as batteries, are suggested to be combined with PV sources to sustain the continuous power supply to the connected loads regardless of the power fluctuations in the PV sources [23]. Additionally, the integration of the grid with hybrid PV-battery system allows for a higher degree of deregulation on the demand side, which may result in achieving lower running costs for high performance.

The objective of this paper is to provide a grid-connected PV system with energy storage device based fast charging solution for transportation infrastructure for high EV penetrations. An efficient configuration of the proposed system using a multiport converter (MPC) and an optimal power flow management tool are both desirable to supply the demanded high-power charging rates. Thus, new converter topologies are presented in this paper, and the novel power flow management proposed could result in a more efficient operation of the system. An overall image of the proposed grid-connected PV system with energy storage devise for the EV charging station is shown in Figure 1. The presented hybrid system is utilizing DG technology by the adoption of the PV/battery sources on the demand side. These DG sources are connected to the electrical grid in a grid-tied system to provide fast charging power rates to the EVs. Further sections of the paper are arranged as follows: Section 2 depicts the charging scenarios available for EVs. Section 3 examines the various PV–EV charging architectures and different converters available for integrating the PV, EV, and ESD with the grid. Section 4 depicts the intelligent energy management and dynamic power flow between the various sources and loads in a grid-connected PV–EV with ESD. Section 5 discusses the methodology for formulating the objective function for power flow optimization using particle swarm optimization (PSO). Section 6 presents the experimental analysis of the proposed system under different modes of operation, and the results depict the efficiency of the proposed control and management techniques.

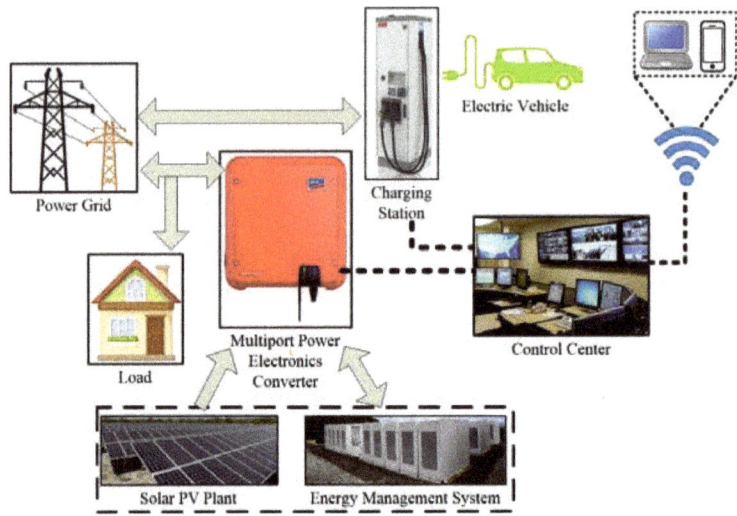

Figure 1. An overview of the proposed hybrid system.

2. EV Charging Scenario

As renewable energy is the future of power generation, it is necessary that it is considered for electric vehicle charging as well. PV panel-based EV charging station is being established in many countries. For understanding the solar-powered EV charging infrastructure [24], it is necessary to realize the existing EV–PV system, which is in use already by the industries or are in the development stage by a different academic institution.

EV Charging

EV charging can be performed by AC [3], as well as DC [25]. Power obtained from the grid is converted into DC for recharging the battery. Range of power required for charging varies for HEV, PHEV, and PEV.

$$P_{cp} = V_{evh} I_{evh}, \tag{1}$$

where P_{cp} denotes Power of DC charge, V_{evh} voltage for electric vehicle charging and I_{evh} represents current required for charging the electric vehicle.

$$E_{cp} = \int_0^{t_{cp}} P_{ce} dt, \tag{2}$$

where E_{cp} is the energy delivered by the batter over t_{cp} period.

When AC based charging is performed on EV, an AC/DC power converter is required for connecting the battery with the 1- or 3-phase system. AC charger used globally can be categorized into three types.

Type 1: It is a single-phase charging method generally adopted by the US (SAE J1772-2009) [26]. Three pin plugs are used for charging (phase, earth and neutral).

Type 2: It is a single, and three-phase charging method generally adopted by Europe (VDE-AR-E-2623-2-2) [27]. Three-phase plugs contain five pins (Phase 1, Phase 2, Phase 3, neutral and earth).

Type 3: It is a single, and three-phase charging method generally adopted by Alliance.

As per IEC 61851-1 standards [28], the charging can be classified into four modes of operation. Both mode 1 and 2 derive from the standard power socket. In the case of mode 2, inbuild protection is provided to the system. Mode 3 consists of electric vehicle supply equipment (EVSE), which ensures protection and control functionality to be present in chargers. Mode 4 discuss the DC-based charging, which is generally accounted when the power of the electric vehicle is more than 50 kW. DC-based charger reduces the requirement of AC/DC converter on board hence reducing the space and weight constraint of onboard chargers. The most commonly used DC charger is a combined charging system (CCS) [29], Type 4 CHAdeMO [30] and Tesla dual charger [31]. The major benefit of using a DC-based charging is that a bidirectional flow can be created between vehicle to home, building, load, grid or any other power deficit system easily.

Economical and sustainability are two of the merit for using PV systems for electric vehicle charging. The implementation of solar energy for charging results in higher efficiency regarding the contacts of fuel usage and the life cycle of the battery [32]. As it can be determined from the discussion above DC-based charging is mostly preferred for PV based EV charging. As both the EV and PV are of DC nature, it's easy to implement, and there is a possibility of smart charging, which indicates the variation of charge concerning time. V2G can also be achieved easily by the implementation of DC charging.

For charging the EVs from solar energy, various architecture can be implemented. EVSE based charging in case of grid connection is implemented for charging EV directly for AC grid. Two of the conventional power converters that can be implemented for the integration of PV, EV, and grid are (i) Multi port-based converters for PV, EV, and grid integration. (ii) Separated power converters for EV, PV, and grid with common point interlink. AC or DC interconnects are required between different power converters. They help in interconnection the PV power amongst different EV and EV power to the grid. From the power converter types mentioned four types of architecture could be deduced for charging of electric vehicles.

Architecture 1: AC interconnection along with separate converter for PV and EV

The separate converter is used for PV panel and EV charging and discharging. PV panel consists of a DC/AC converter with maximum power point tracking (MPPT), whereas the AC/DC converter is implemented at the EV end. A grid of 50 Hz has a major role to play in this architecture as all the power is passed through the grid. The major drawback is that DC power generated by PV can't be directly used for EV charging, creating an unnecessary requirement of DC to AC and AC to DC conversion in the process.

Architecture 2: DC interconnection along with separate converter for PV and EV

In this architecture, both PV and EV requires a DC/DC converter. For EV charge control needed to be present with converter, whereas, for PV, an MPPT controller is required. DC interconnection helps in the utilization of DC power from the PV for battery charging directly. For grid interconnection, a central inverter is used. The central inverter is vital as it draws power from PV and EV depending on the demand side requirement. Separate construction of DC interconnection makes the architecture less desirable as the existing AC base infrastructure is not utilized to its full extent.

Architecture 3: AC interconnection along with multiport converter for PV and EV

In this architecture, the PV–EV and grid are interconnected to a central DC link using multiport converters. With the help of grid multiple, MPC relates to each other. Because of the interconnection control of the system becomes easy along with low effective cost and high-power density. The previous architecture used communication-based control, whereas in this architecture, control is provided directly via MPC. One of the major disadvantages of this system is that DC for PV using one MPC can't be used to charge the EV connected to alternative MPC without AC conversion.

Architecture 4: DC interconnection along with multiport converter for PV and EV

This architecture consists of merit from architecture 2 and 3. MPC is used for interconnection of PV and EV. DC interconnection is used to link different MPC. The grid is connected to the system with the help of a high-power central inverter. All the architectures corresponding to charging of EV from PV are depicted in Figure 2.

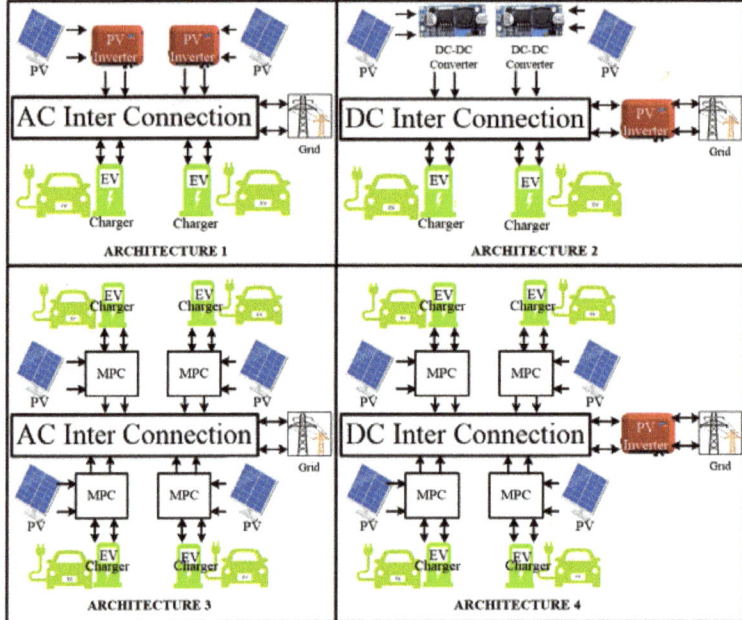

Figure 2. System architectures for electric vehicle (EV)—photovoltaic (PV) charging using multiport converters.

3. PV Assisted EV Charging

This section examines the design of charging infrastructure for an electric vehicle with the help of PV panels. The proposed system can be adapted for residential areas and workplaces to charge electric cars as per the convenience of the EV user. The major objective here is to maximize the use of PV energy for EV charging by utilizing energy storage systems and minimize the energy exchange with the grid.

3.1. System Design Architectures for Solar-Powered Charging Stations

Messenger [33], and Stapleton [34] suggest that for any PV based system to operate effectively, the entire system design should be done by considering the parametric and location constraints for installation of PV systems, load requirements, and electric codes as per the location. Considering these constraints, there are three major types of system design for charging stations: Off-grid PV system with energy storage device (ESD), grid-connected PV (GCPV) system with ESD, and GCPV system without ESD. Conventionally, an off-grid PV system with ESD type of architecture is used for EV charging stations. In this architecture, energy from the PV module and, ESD (battery) together is used for meeting the load power requirements, and any excess power is fed back into ESD. The major drawback with this system is that effective utilization of this type of architecture can be only done if an external control system is incorporated. Apart from the control aspect, the system is not reliable, due to its dependency on solar irradiance [35]. In order to overcome these drawbacks, charging stations associated with the GCPV System without ESD were implemented. In this type of architecture, the PV power generation is mainly used for meeting the load power requirements and the excess power other than charging EV load is injected into the grid. The grid acts as a storage unit in this type of architecture. The only disadvantage of this system is, loss of energy storage capabilities with grid failure could pose a difficulty to meet load requirements.

Considering all these concerns and constraints, the GCPV System with ESD is developed in this research. The basic structure of this system is shown in Figure 1. The major advantage with this system is, it is more reliable when compared to the conventional systems like this it will continue to satisfy loads even when the grid fails. Also, it can be designed to have less reliance on the grid.

Apart from it, 100% utilization of time to use (TOU) pricing is possible in this type of system with the presence of an additional controller to control the power injected into the grid from the DC side.

3.2. Power Converter Types for Grid Connected PV with ESD Assisted EV Charging

A grid-connected PV with energy storage device assisted EV charging system would involve the PV system, AC grid, energy storage device, EV and a power electronic converter interface to combine and link them. This converter must be capable of enabling the charging of EV from all the three utilities, i.e., PV and AC grid and facilitate charging of EVs and ESD from both the sources. This operation will guarantee that the grid can support the EV through G2V during the absence of PV and ESD and can charge the ESD for emergency power requirements. In this section, an optimal multiport converter topology that integrates EV, PV and the AC grid are proposed. The need to design such an optimal topology is necessary to ensure that the converter has low cost, high power density, and high efficiency. Conventionally, an optimal converter topology is chosen by considering several parameters, such as converter volume, efficiency, the number of components, controllability, ripples, and opportunity for efficiency improvement.

3.2.1. DC Link-Based Converter

Considering the above parameters, initially, the converter types are designed based on a DC-link. These types of systems act as a high voltage energy buffer among the ports. Conventionally, there are 3 sub converters in a DC-link based multiport converter with multiple converter topologies. These control strategies include a boost converter and MPPT algorithm integrated with the PV array for maximum power, a buck-boost converter for controlling the charging needs of EV and ESD. The complete setup is linked to the local grid with the help of a voltage source converter, which operates bidirectionally and which is accountable for power stability with the grid. The major advantage with such topology is, it's simple power flow between PV and EV, and minimized DC/AC conversion losses. This topology also improves the AC/DC conversion of VSC for charging the EVs during peak load and when both PV and ESD were exhausted. Apart from the advantages mentioned, the topology possesses some major drawbacks, due to its large DC link capacitors and multiple controllers for each sub-converter.

3.2.2. Impedance-Network Based

Carli [35], Rasinab [36], and Rasin [37] discussed impedance network based multiport converter topology. These converters deal with a variable DC-link voltage connected with a secluded DC-DC converter for EV charging. These converters have major advantages, due to their low component count, which improves the reliability of the system. The converter topology also has an advantage over fewer control algorithms. The major drawback for such type of converters is that it is most intrinsically modular and has very high control complexity, due to variable DC link voltage. The system architecture of both DC link based multiport converter and impedance network-based converter is depicted in Figure 3a,b.

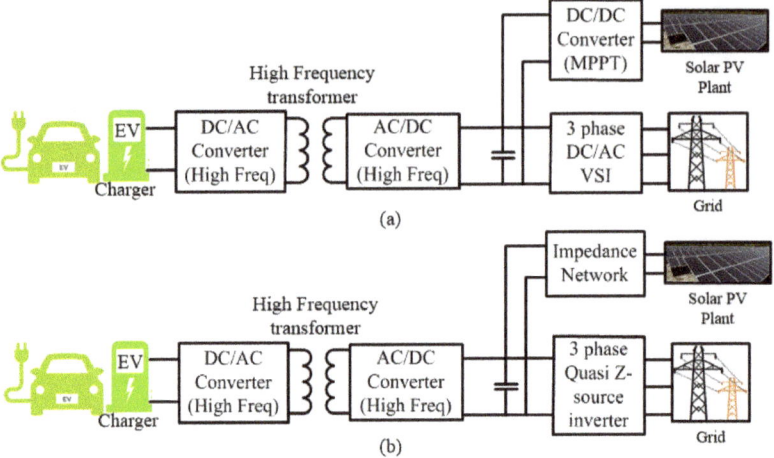

Figure 3. Block representation of grid-connected PV–EV converters (**a**) DC-link based converter, (**b**) Impedance-network based converter.

3.3. Multiport Converter

For this research, the DC link based multiport converters are used with modified control and intelligent energy management system. There are two types of multiport converters, as shown in Figure 4.

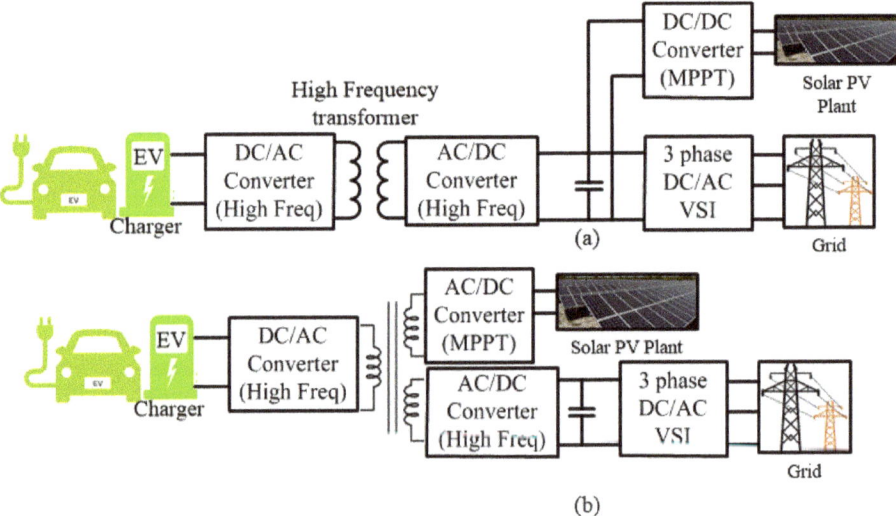

Figure 4. Block representation of multiport converters. (**a**) MPCA; (**b**) MPCB.

The major difference between both the possible MPC's is the capability of isolating PV panels from the grid. The Figure 4 depicts that, both the topologies adapt a central DC-link available within the DC-DC converter. For efficient power flow management, the DC link voltage rating must be higher than the peak voltage of the system. In Figure 4a, a non-isolated DC/DC converter is adapted for achieving maximum power from the PV array. A high-frequency DC-DC conversion is adapted for EV charging station by adding a high-frequency transformer between the conversion process as per IEC 61727. For this research, the converter topology in Figure 4a is considered and tested with the proposed energy management strategy.

4. Power Flow Management

In order to design the proposed system configuration, the flow of the power between the four main elements in this system needs to be explored. The main elements are the connecting electrical grid, the PV sources, the battery storage and the EVs charging load. The decision on the need for a bi-directional power flow power electronic system, along with their sizing requirements, can be decided based on power flow management. Consequently, the research attempts to solve the power flow management problem by introducing their applicability on the studied application.

Conventionally, the power flow in a grid-connected PV/battery system is predefined by heuristic rules that consider the load demand, the PV insolation levels and the off-peak utility hours [38]. However, a dynamic grid tariff complicates the solution of the proposed system further. In a dynamic grid tariff system, the operation of the PV/battery system using the simplified heuristic rules will provide running cost solutions that largely deviate from the minimal cost operation. Thus, the research in this area has taken on an accelerated path.

In Reference [39], a Lagrangian relaxation technique is applied to determine the optimal hourly battery charging or discharging current, where the objective is to maximize the contribution of the hybrid system to the grid. The proposed technique in Reference [39] assumes there is no dispatch cost associated with the PV/Battery output power. This leads to the negligence of the battery degradation cost and its advisable operating conditions. Additionally, the formulation of the problem is limited to a thermally based electrical grid system. A predefined rule-based model is presented in Reference [40] where the battery energy storage is integrated into the renewable energy system in order to enable the PV source to act as a dispatchable unit on an hourly basis. The objective of the battery storage utilization in this paper leads to solutions that are not necessarily minimizing the running cost. Additionally, the system is sensitive to solar power forecasts.

From the above literature, it is observed that, initially, the problem formulation should account for the aging factor of the battery in order to extend the battery lifetime, and thus, increase the system reliability. The desired power flow management topology has to accommodate non-linear functions. This allows the generalization of the developed topology on different operating scenarios. An online error compensation stage has to be included in the topology to allow the system to operate effectively at mismatching conditions and forecasting deviations. Lastly, the online optimization stage should be designed to operate with low computational time, which makes it easily integrated into real-time controllers. Study regarding intelligent energy management strategy and dynamic power allocation is required in order to overcome these drawbacks and achieve the proposed optimization.

4.1. Intelligent Energy Management Strategy

The EV–PV charger in its present structure will probably charge the EV from sun-based energy, yet it does not possess any knowledge of its own. The energy price will remain low in early hours of the day as per the forecasting; hence morning period is advantageous to charge EV from the grid. While sunny afternoons provide advantage for charging from the solar. For the realization of control of EV–PV system, smart charging algorithm necessary.

The situation of charging the vehicles at parking stations has been considered where every vehicle has a predictable time of accessibility as load. The user defines the charge time window as input while parking the vehicle, and the other preferences of the user could be pricing, type of charge, etc.

The other characterization of the vehicles is its capacity, chemistry, open circuit voltage level, state of charge, temperature, etc. Depending upon the user preference and battery properties every vehicle will assign the power to the system. Hence to conclude, energy utilization of the system is optimized by considering customer preferences and load attributes using dynamic power management for loads. The control techniques, advanced demand-side power management and clients' preferences in utility services are the main objectives for grid design. For that purpose, a given system has been designed to show the elements of the smart grid. The basic structure of the complete system is shown in Figure 5. This system accomplishes all the merits discussed and designed using energy management system, power grid, charging and battery loads. The iEMS used to control every parking deck, and these parking decks are made of multiple loads.

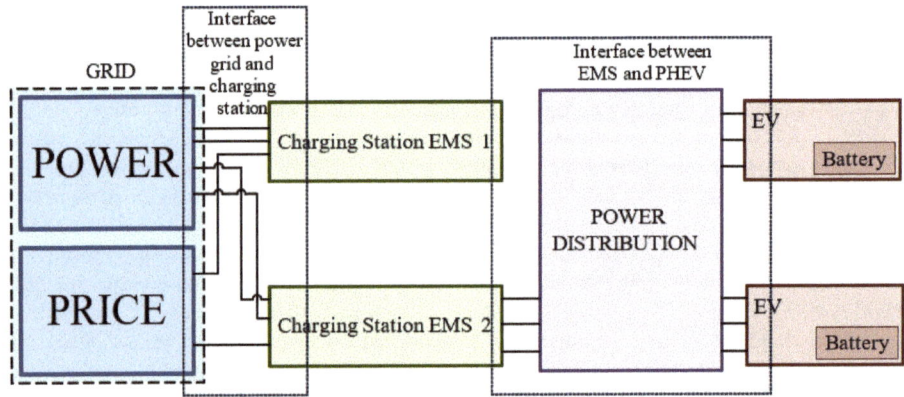

Figure 5. Architecture of the intelligent energy management system.

All the information related to the power available and the pricing as per the time period at the grid side is automatically updated to the iEMS. This process helps in simultaneously monitoring the multiple converters for making real-time decisions by iEMS.

4.2. Dynamic Power Allocation for Energy Management at EV Charging Station

The primary components of the system are EVs, ESD, energy management systems and the power grid (utility). We consider vehicles clustered in parking decks, with each parking deck under the control of an iEMS. Essentially, a cluster of vehicles and an energy storage device being controlled by an iEMS is a manageable load for the grid. The iEMS acts as an interface between the vehicles and the utility. Information about available power and pricing is regularly updated to the iEMS; multiple chargers at a parking deck are then simultaneously supervised and controlled according to the real-time inferences made by it.

At the time of plug-in, the customer will specify their preferences on the type of charge desired, estimated time of availability, willingness to participate in V2G, the price they are willing to pay for charging, etc. This information may be entered via an interface within the vehicle itself or an external user GUI available at the parking deck or from an internet profile maintained by the user for that vehicle [41]. The parking deck can be equipped with an intelligent charger at each vehicle parking space, capable of acquiring the vehicle battery state and relaying it to the iEMS via a communication medium; alternatively, each EV could be communication enabled, capable of sending its battery data to the iEMS.

The electric load that can be sustained by the utility from a parking lot will change during the course of the day as the system load varies. For vehicles willing to provide electricity back to the grid, opportunities could also arise for V2G operation depending on the grid requirements. Thus, the varying price of electricity can also be incorporated into the system information for better decision making. Using the information from the utility and the vehicles, the iEMS will make a real-time decision on power sharing to each vehicle and communicate this to the intelligent chargers.

Since the state of the system is continually changing with the arrival and departure of vehicles and changing power requirements from the utility, the system states are sampled at regular time intervals as the power allocation needs to be recomputed each time a vehicle plugs in/out, or there is a change in power availability. The system is hybrid in nature; its event-based character arises from the plug-in, plug-out activity and sampling time steps used by the iEMS for making its decisions. Moreover, the process of vehicle charging is continuous with a non-linear power consumption curve, which varies dynamically. The randomness of the preliminary states of charge, plug-in and out times and varying power curves introduce dynamicity in the system, which makes the optimization for power allocation a large scale, nonlinear, time-varying multi-objective problem with multiple constraints. In the ensuing sections, we provide a framework for system modeling and operation, addressing system goals and constraints along with optimization on a chosen objective.

4.2.1. System Modeling

The interactions in the system occur between the iEMS and the utility on the one hand and the loads (vehicles) and the iEMS on the other. The iEMS acts as broker agent, taking information about the power available from the utility on the one hand (can also be the transformer rating at the parking deck), and the vehicle battery states and user preferences on the other hand. It makes a decision on power allocation which is optimal in terms of satisfying both the parties. The system components can be modeled on 'Agents Based Approach' [42], in which each entity has a set of attributes, states, and functions and the entities interact with each other in order to achieve individual and system goals.

4.2.2. Optimization Objectives

A number of objectives are possible for the problem. Many objectives can be formulated around user preferences. For example, the minimization of the time taken to charge the vehicle battery for all the users, according to the price they are willing to pay. For users that do not mind compromising on the time and would like to charge at the lowest prices, the objective could be to only charge when the electricity cost is below a threshold. If the vehicle to the grid is also activated when the power flow could be facilitated such that profit to each user is maximized [43]. Another objective could be to minimize the overall power consumption (if desired by utility) while trying to guarantee a minimum threshold SoC (say, 60%) for each vehicle. In this paper an optimization problem is formulated along with the required constraints and the operating cost function is chosen as a combination of electricity grid prices and the battery degradation cost. The proposed optimization procedure uses PSO, which acts as a prediction layer for forecasting the system operation. This data is further processed for switching between the different sources as per the load requirement and power available.

4.2.3. System Constraints

The primary constraint considered here is the power available from the utility. Type of charge (slow, medium, fast), time of availability in the parking lot, the minimum desired state of charge at plug-out, maximum price the user is willing to pay for charging, the maximum power that can be absorbed by a vehicle battery, other battery requirements, etc. are other possible constraints. The user could also specify the number of miles he/she plans to drive after plug out, in which case, the SoC for achieving this will be guaranteed at plug-out. Another constraint for the iEMS could be in terms of the layout of the parking deck, the sizing and capacity of the cables will place a limit on the power that can be channeled to each vehicle(s). For a more robust system, we could consider the abrupt plug-out by a user before the stated time, and in this case, the aim would be having some fairness in the SoC distribution at every time step to have a reasonable SoC even before plug-out. Additional constraints in the system could be in terms of the bandwidth availability of the communication channel for sampling the states of the vehicles, which would limit the sampling time. System performance with packet delays and drops could also be evaluated. The final goal of the iEMS is to be flexible in terms of accommodating multiple objectives for different users. It also opens up the opportunity of dividing the station into clusters, grouping vehicles/users with common objectives and assigning the responsibility of optimizing on each cluster to a sub-iEMS thereby translating the problem into the arena of distributed control. The decision on how much power to allocate to each sub-iEMS will be made by the central iEMS or could also result from bidding actions by the sub-iEMSs.

4.3. Particle Swarm Optimization

PSO is an iterative stochastic optimization algorithm, which is derived from studying the pattern in which a flock of birds or schools of fish travels [44]. Multi-dimension solution space is searched by the algorithm by collective search containing different particles and the best solution found by the other particles are communicated. The communication enables the system to take an informed a decision about the movement of each particle to find the best solution available global. Weighting factor and random variations are also considered in the algorithm to prevent any early convergence whenever local minimum is present.

Following the analysis in Reference [45], the electric vehicle system constraints are adapted from Reference [46]. Each constraint is defined by the user in the EV-iEMS. A given day is fragmented up into different intervals as per the irradiance variation pattern.

5. Methodology

The initial configuration of the grid-connected PV system with an energy storage device, explored in this paper is shown in Figure 1. The PV array, energy storage, and EV charging station (EV load) are connected at the DC link through a high frequency DC/DC converter. An interconnection between the DC link and the electric grid is achieved using a bi-directional AC/DC converter. The DC/DC converter used at the energy storage is a bi-directional power electronic interfacing converter to allows smooth operation during the battery charging and discharging modes. However, the uniqueness of this study is the reliance on the proposed configuration on the power flow management results. Thus, the shown configuration will be developed at the end of this section to reflect the findings of the optimal power flow operation through a given period of operation. The battery power (P_b) is considered to be negative while it is charging and positive during discharging. All the charging slots, and additional loads connected with the system are summed up to form the load power (P_L). Here, the load power (P_L) is considered positive in all the cases as neither the battery storage, nor the EV supply power back to the grid. The grid power (P_G) is negative while the distributed generation feeds power into the grid, and it is considered positive when the grid is satisfying the loads.

Power flow management methodology is developed by considering all the operating modes and assumptions. Initially, the forecasting data regarding the weather conditions in the area, power generation both from PV and grid, grid tariff and the power consumption profiles of auxiliary loads and the EV charging loads are considered for a particular time period. The process of determining the forecasted data and the grid tariffs were thoroughly discussed in the literature and can be applicable [47,48].

Once the forecasted data and the grid tariffs were obtained, the objective function is defined. The defined objective function aims at optimal power scheduling between the power sources and energy storage for minimizing the daily running cost of the system. In this paper PSO method, as discussed in this section, is used to define the optimization problem. In this paper, the implementation of PSO for power flow optimization is considered to be a predictive upper-level optimization stage, where the process is based on early forecasted data. The system measurements considered for this methodology are, PV output power (P_{PV}), grid power (P_G), load power (P_L), SOC of ESD and EVs, and grid tariff (G_T).

The power balancing equation for GCPVS with ESD is given by Equation (3), and the state of charge of the ESD and EVs is estimated using Equation (4):

$$P_b(t) = P_L(t) - P_{pv}(t) - P_G(t), \tag{3}$$

$$SOC(t) = SOC(t - \Delta t) - \frac{P_b(t) \Delta t}{Q}, \tag{4}$$

where Δt is the time interval, and Q is the SOC of battery at the time of power exchange.

The system operation is constrained by the following limits.

$$SOC_{min} \leq SOC(t) \leq SOC_{max}, \tag{5}$$

$$P_{b_{min}} \leq P_b(t) \leq P_{b_{max}}, \tag{6}$$

$$P_{G_{min}} \leq P_G(t) \leq P_{G_{max}}. \tag{7}$$

The system constraints defined in Equations (5) and (6) are considered for improving the battery lifetime, while Equation (7) defines the limits for utilizing power from the grid. At any given time (t), the total cost of the system is expressed as:

$$C_T(t) = C_G(t) + C_{BD}(t), \tag{8}$$

where $C_T(t)$ is the total cost of the system, $C_G(t)$ is the operating cost of grid $C_{BD}(t)$ is the degradation cost of the battery.

By considering the above equations, the system objective function for operating the system for a given time period is given by Equation (9).

$$Min(C_T) = Min \sum_{t=0}^{t=T} [C_G(t) + C_{BD}(t)], \tag{9}$$

where the grid operating cost $C_G(t)$ is given in Equation (10):

$$C_G(t) = GT(t) P_G(t) \Delta t. \tag{10}$$

However, the process of developing the degradation cost function for the energy storage device is a complex process it involves various independent factors [49]. The three major aspects impacting the operation, lifetime and health of the ESDs are battery temperature [50], average SOC [51], and the depth of discharge (DOD) [52].

By considering the temperature impact, both the charging and discharging process affect the temperatures of the ESD. This phenomenon directly affects the average power (P_{avg}) drawn from or to the battery as given in Equation (11). In order to account for temperature variations in both the charging and the discharging scenarios the absolute of average temperature is considered in the equation.

$$T = T_{amb} + R_{th} |P_{avg}|, \tag{11}$$

where T is operating temperature of the battery, T_{amb} is the battery pack ambient temperature, R_{Th} is the thermal resistance of the battery pack.

The effect of battery temperature on the lifetime of the battery is calculated from the Arrhenius equation [53].

$$L(T) = aQe^{\frac{bQ}{T}}, \tag{12}$$

where a and b correspond to the curve fitting parameters.

The impact of temperature on the battery degradation cost is given by Equation (13)

$$C_{temp} = C_{bat} \int_{ti}^{tf} \frac{dt}{T_i \, L(T)}, \tag{13}$$

where T_i is the number of intervals in the total time period, and C_{bat} is the initial cost of the battery.

Apart from the temperature impact, average SOC of the battery also impacts the degradation rate of the battery. A firm relation between the average SOC and battery degradation cost is given by Equation (14):

$$C_{soc} = C_{bat} \frac{m \, SOC_{avg} - d}{Q_{fade} \, n \, T_i}, \tag{14}$$

where m, d and n are constants of curve fitting, and Q_{fade} is the capacity fade at the end of the battery lifetime.

Finally, a relation between the lifetime of the energy storage device and the DOD of the battery are presented in References [54,55]. Equation (15) depicts the cost per kWh of the battery DOD.

$$C_{DOD} = \frac{C_{bat}}{2 \, N \, DOD \, Q \, \mu^2}, \tag{15}$$

where N is the battery lifecycle for a particular DOD, and μ depicts the charging and discharging efficiency of the battery.

Considering all the factors that impact the degradation of the battery, the cost function for battery degradation is given by Equation (16) and the final objective function is achieved using (17) after discretization,

$$C_{bd} = \max\{C_{temp}, C_{DOD}, C_{SOC_{avg}}\}, \tag{16}$$

$$C_{T_i, \, run} = \sum_{k=1}^{N_i} [C_G(k) + P_b(k) \Delta t C_{bd}(k)]. \tag{17}$$

The proposed methodology for optimizing the power flow contributes to the improvement of the lifetime of the system components, especially the storage devices. The PSO implementation for the optimal selection of different power sources is illustrated in Figure 6.

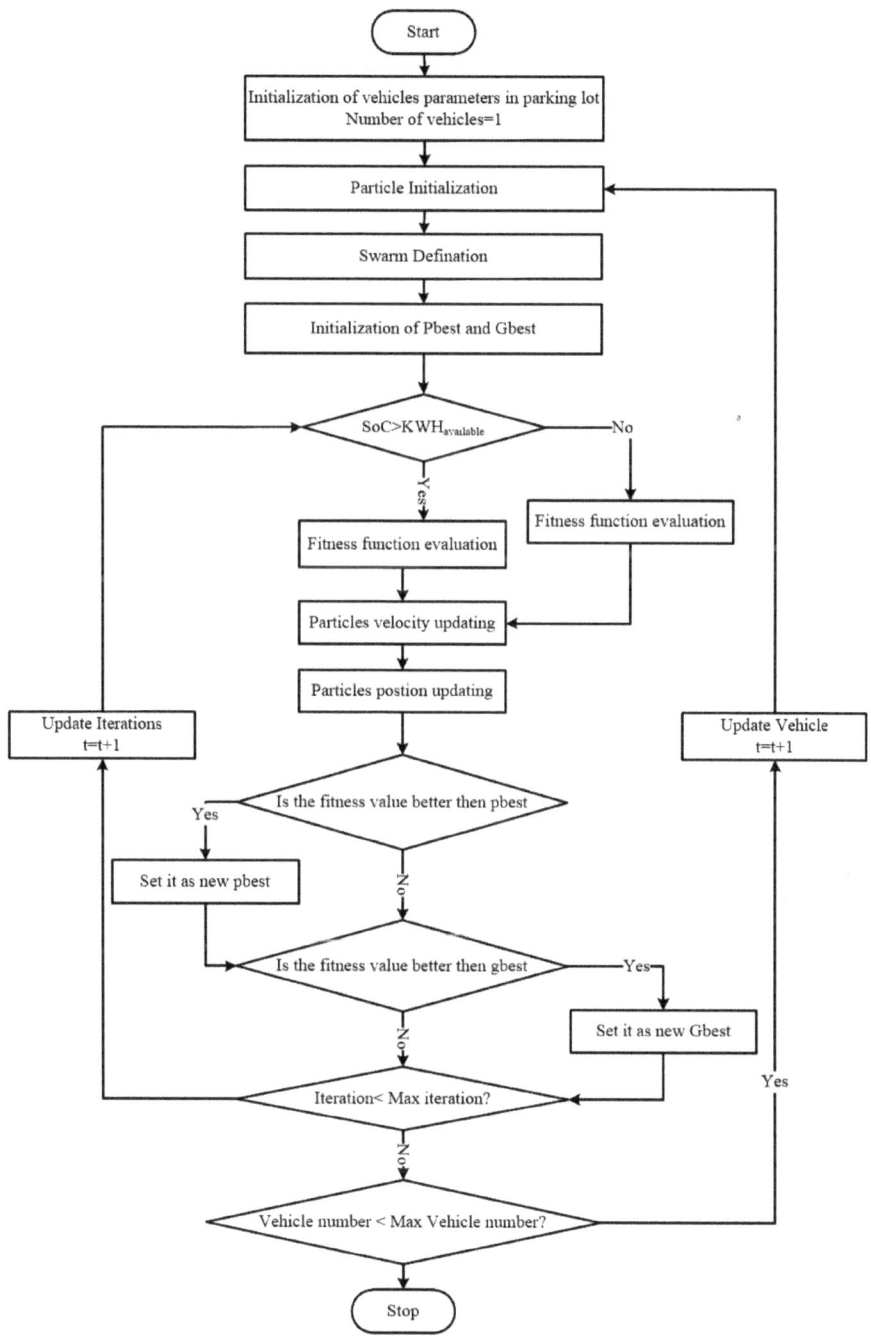

Figure 6. Implementation of particle swarm optimization (PSO) for the optimal selection of power source.

6. Experimental Work

A simulation system model, containing the realistic models for the electric network and the EV, was developed to test the developed charging strategy and study its effects on the electric network. A simplified illustration of the whole simulation system is presented in Figure 7.

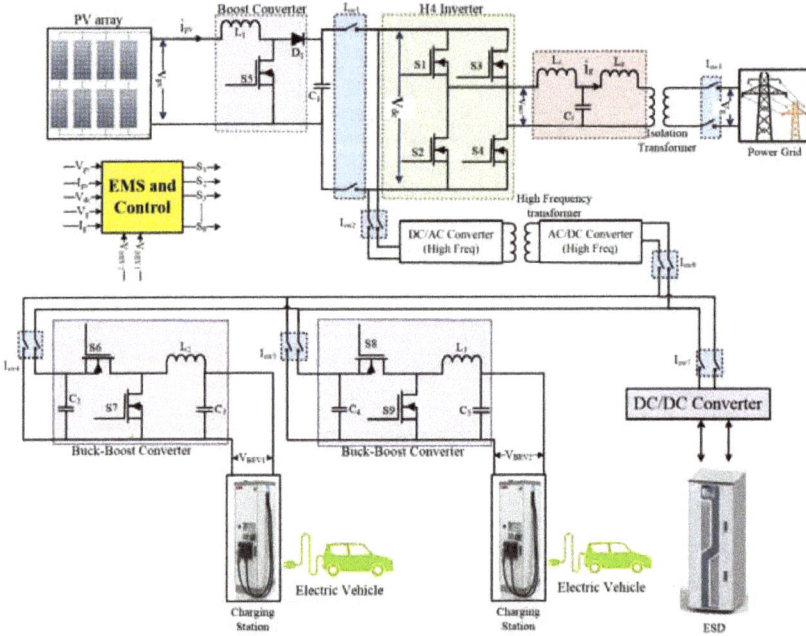

Figure 7. Simulink implementation of grid-connected PV–EV (GCPE) charging architecture with energy storage device (ESD) and intelligent energy management system (iEMS).

These models are developed in the MATLAB/Simulink environment. The main script takes data on driving distances and solar irradiance as input parameters. It then calculates the amount of power that should be supplied to the EVs and the electricity grid respectively each hour, in order for the self-consumption to be maximized and grid impact to be reduced. The parameters used for the simulation of the proposed system is given in Table 1.

Table 1. Simulation parameters of grid-connected PV–EV charging station with an energy storage unit.

Components	Rating
Solar Array	20 kW
Boost Converter	5 kHz, 500 V
DC link Voltage	500 V
Grid Parameters	2500 MVA, 120 kV
High Frequency Transformer	25 kva, 5 kHz, 500:80:80, three winding transformer
Electric Vehicle	Mahindra severity
EV Battery	Lithium Battery Nominal Voltage = 72 V Rated Capacity = 200 Ah
Energy Storage Device	Rated Power = 150 kW Power Conversion Efficiency = 90%

6.1. Electric Network Model

The used electric network model represents a typical grid-connected PV system with an energy storage unit. The details for this synthetic electric network model are taken from MathWorks [56]. The schematic of the developed electric network is presented in Figure 7. The developed system consists of a 20 kW solar array connected to a 5 kHz boost converter with incremental conductance MPPT algorithm. The PV network is connected to the grid via a three-phase voltage source converter (VSC). The VSC can be operated both in inverter and rectifier mode in order to facilitate the power flow both from the grid to charging stations and from PV to the grid. A 25-kVA high-frequency transformer is used to provide isolation between GCPV and the EV charging, ESD unit.

6.2. EV Charging Model

The EVs can be modeled as systems composed of many dynamic dependencies. The key components in the EV models are the charger and the battery. The EV model used in this research is sourced from Reference [57]. It consists of blocks for the driving schedules, charger logic, converter losses and the battery model as presented in Figure 8. The driving schedule block collects and passes on the data regarding the EV ID, EV address and the EV demand $DoD_{EV,n}$. The EV demand describes the Depth of Discharge (DoD) of the nth EV. The EV ID specifies which of the EVs are connected and the address specifies the household node for each EV. As soon as an EV is connected to the electric network, the driving schedule block relays the details for the charger logic block and updates the battery DoD information to the battery model block.

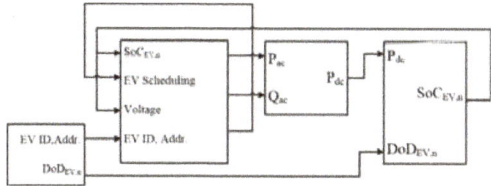

Figure 8. Architecture of EV charging model.

The charger logic block receives data from four sources. The first input comes from the battery model block, which continuously updates the charger logic block with the current $SoC_{EV,n}$. The second input comes from the EV fleet aggregator who submits the EV schedules. These schedules are directly used by the charger logic block. The third input receives the current voltages from the electric network model. The fourth input comes from the driving schedule. The charger's main function is to convert the AC power into DC power, which is essential to charge the battery. The charger logic block processes the received data and regulates the battery charging in a manner that the local network voltage limits, set for each EV separately, are not violated. It also computes the appropriate charging power rates as they are dependent on the batteries' $SoC_{EV,n}$ levels, as explained in Reference [57].

6.3. Modes of Operation

The model is built on the idea of matching hourly charging powers and PV power generation, meeting daily charging demands and achieving constant power exchange with the electricity grid. Since the driving distance for each EV is randomly selected each day, the charging demand of the car park is unique for each day of the simulated year. The level of self-sufficiency depends on the amount of installed PV power. If the accumulated PV energy generation for one day equals the car park demand, all PV generated power is supplied directly to the EVs. If the accumulated PV energy generation falls short of the daily car park demand, the deficit is supplied from the electricity grid, evenly distributed over all workday hours. If the daily PV energy generation exceeds the car park demand; the excess power is supplied to the electricity grid as evenly distributed as possible. In this case, where the PV system generates excess energy, the power exchange with the electricity grid is different each hour, depending on whether the hourly PV energy generation is higher or lower than the mean excess energy generation. The complete scenario of charging the EVS by considering all the constraints is explained in nine different modes, as shown in Figure 9.

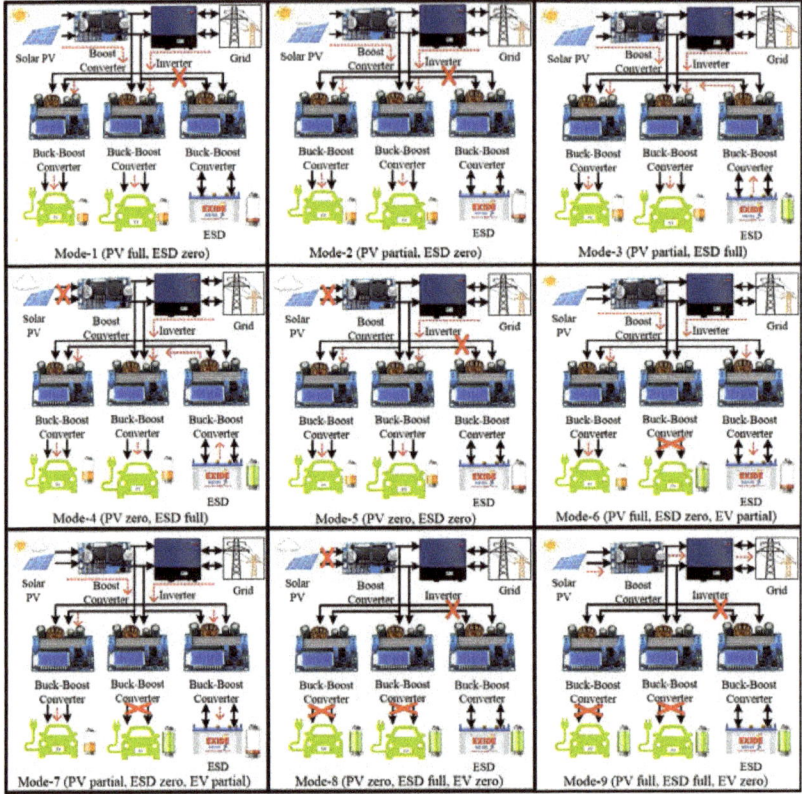

Figure 9. Modes of operation for GCPE charging architecture with an ESD and iEMS.

Mode 1: PV generation full, EV load Available and ESD has less charge.

In this scenario, the PV generation is happening under standard test conditions and considered to generate maximum power. Since all the systems EV1, EV2 and ESD act as loads and are ready to charge, and the power demanded is greater than the power generated from the PV. In this condition, the grid supports the PV in satisfying the load.

Mode 2: PV generation partial, EV load Available and ESD has less charge.

In this scenario, the PV generation is considered to happen under varying irradiance conditions. Here the power generated from the PV is low. Since all the systems EV1, EV2 and ESD act as loads and are ready to charge, and the power demanded is greater than the power generated from the PV. In this condition, the grid supports the PV in satisfying the load.

Mode 3: PV generation partial, EV load Available and ESD is fully charged.

In this scenario, the PV generation is considered to happen under varying irradiance conditions. Here the power generated from the PV is low. Since both EV1 and EV2 act as loads and are ready to charge, and the power demanded is greater than the power generated from the PV. In this condition, both the grid and ESD supports the PV in satisfying the load depending upon the requirement of the user.

Mode 4: *No PV generation, EV load Available and ESD is fully charged.*

In this scenario, the PV generation is zero. Since both the EVs are ready to charge, and the power demanded is greater than the power generated from the PV and power available from the battery. In this condition, the grid supports the ESD in satisfying the load and PV remains disconnected from the system.

Mode 5: *No PV generation, EV load Available and ESD has less charge.*

In this scenario, the grid satisfies the changing needs of the EV station. The PV remains disconnected from the charging dock, and once the charging needs of the EVs are satisfied, and the load connected to the utility is less, then the grid charges the ESD.

Mode 6: *PV generation full, Partial EV load Available and ESD has less charge.*

In this scenario, the PV generation is happening under standard test conditions and considered to generate maximum power. Since the ESD has less charge, and EV load available is less and ready to charge, the PV satisfies the EV load first and then charges the ESD.

Mode 7: *PV generation partial, Partial EV load Available and ESD has less charge.*

In this scenario, the PV generation is considered to happen under varying irradiance conditions. Here the power generated from the PV is low. Since the ESD has less charge, EV load available is less and ready to charge, and the power demanded is greater than the power generated from the PV. In this condition, PV satisfies the EV load first and then charges the ESD depending upon the generating capability of the PV.

Mode 8: *No PV generation, No EV load Available and ESD is fully charged.*

In this scenario, the grid satisfies the utilities connected to it. The ESD remains connected with the charging docks.

Mode 9: *PV generation full, No EV load Available and ESD is fully charged.*

In this scenario, the PV generation is happening under standard test conditions and considered to generate maximum power. Since the EV load is not available and the ESD is fully charged, the PV transfers its power to the grid.

By bearing these scenarios in mind, the model, considered above, is simulated to observe the efficiency of the proposed charging and converter control strategy. The optimal power flow is achieved by adapting the cost function achieved in Equation (17). The forecasting data to perform the experiment is considered as follows:

The irradiance profile for performing the experiment is considered as depicted in Figure 10.

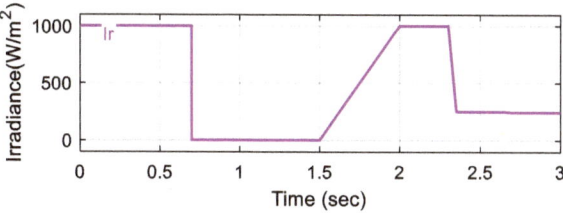

Figure 10. Irradiance variation for PV array.

From Figure 10, it is observed that there is variation in irradiance at multiple instances, this is considered to replicate the behavior of PV system in real-time. By considering the real-time operation, the PV systems depict five times the difference in energy yield for change in environmental conditions. Due to this change, the ratings of the PV converter can be resized by 30%, which corresponds to a 3.2% loss of energy [58]. Apart from the change in irradiance, the impact of using two stage conversion for grid-connected PV system, the switching frequencies, the high frequency DC-DC converter at the charging end and the impact of varying loads in the system also contribute for the power loss in the system. A detailed explanation on power loss in grid-connected PV system and the impact of EV/ESD charging and discharging in a grid-connected system were given in References [59–61]. Since the system design considered for this research investigates the best design for grid-connected PV system to achieve efficient charging of EV and ESD, and the power flow in the considered to be unidirectional, the power loss calculation is neglected.

Apart from the irradiance profile, the major assumptions considered for this research are as follow: The SOC of ESD is 10%and it is considered to be in charging mode and hence it acts as a load, and the SOC of EV1 is 40%, and SOC of EV 2 is 30% and were connected with the GCPV. The process of implementing PSO for developing the proposed methodology as per the forecasted data is given in the following steps.

Initialization of PSO:

Objective: For charging the vehicle finding an optimal power source using Equation (17).

Topology: Swarm optimization star configuration is used to ensure that all particle communicates with each other.

Definition of particles: Depending upon the power capacity of PV and ESD, particles are defined.

Fitness Function: The equation is used to maximize and minimize the power generated during the charging and discharging process.

Search Space: (Constraints) The major constraints to be considered for this problem is that the hours must be a positive real number, which is limited by arrival and departure time.

PSO parameters: The PSO parameters tested for this control are depicted in Table 2.

Table 2. Tested PSO parameters.

Parameters	Value
No of particles in swarm	24
No of iterations	30
Parameters to be identified	2
Search Space Range	[0 50; 0 10];
Swarm declaration	Zero
Velocity Clamping	[3; 1];

The best performance of the proposed switching strategy as per the power available is obtained from the PSO implemented and the corresponding results are depicted in Figure 11.

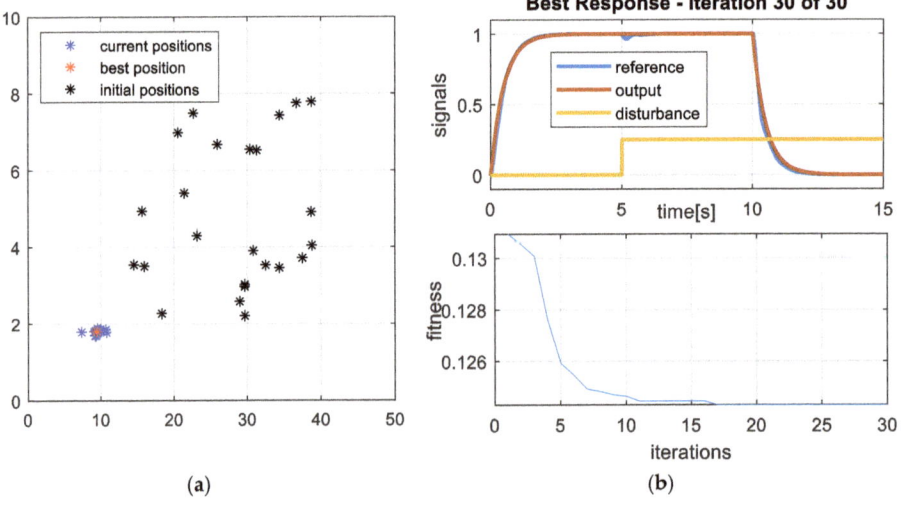

Figure 11. Best switching strategy as per the power available. (**a**) Best position for feeding the load; (**b**) Best response of the algorithm.

Figure 11a shows the best position for feeding the load continuously as per the power available from different sources. The initial position defines the power generation capacity of various sources at the time of interaction with the load. Whereas, the current position defines the source feeding the load irrespective of the availability of the source or the status of the load, and the best position is defined based on the load power and generation capacity. Figure 11b depicts the best response of the PSO in finding the optimal power flow path for satisfying the loads connected at the DC link.

Considering the optimal power flow path defined by the PSO for the forecasted data, the experimental analysis is carried out as follows:

Initially, the PV generation is set to high by operating it under STC. During this period the PV is able to charge the EVs with rated power without any support from the grid. After 0.6 s, the PV generation is set to zero by varying the irradiance of the system. Here the grid comes into the scenario, in order to support the EV charging. At 2.4 s, the PV generation is minimized to generate 20% of the rated power by varying the irradiance of the system, as shown in Figure 10. In this case, the grid continuously supports the charging process of the system. The PSO here monitors the generation and load profile of both the grid and EV charging docks in order to provide fast switching between the sources as depicted in Figure 12.

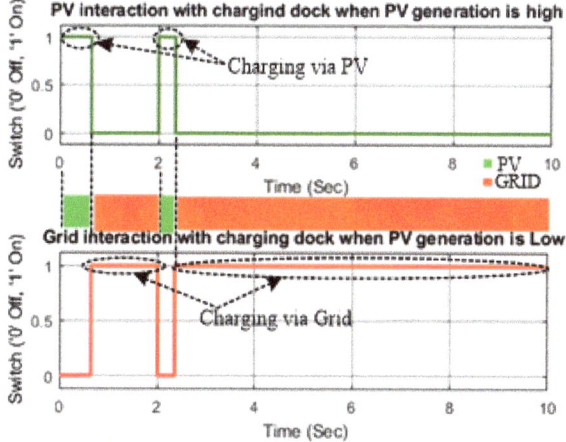

Figure 12. Availability of PV and grid for charging of EVs.

The voltage at the DC link is maintained constant around 500 V, as shown in Figure 13.

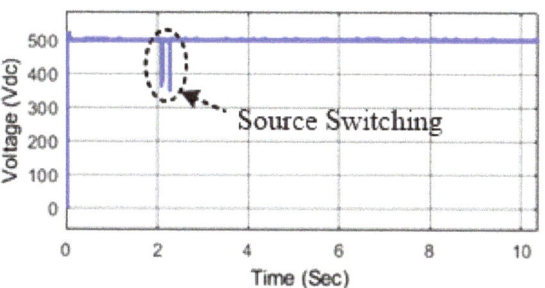

Figure 13. DC link voltage of grid-connected PV system.

A high-frequency transformer is used to provide isolation between charging docks, ESD and the grid-connected PV system. Finally, as per the specification of the vehicle mentioned in the section, the EVs were able to charge at rated voltage and current even with disturbances in PV generation, as shown in Figure 14a,b.

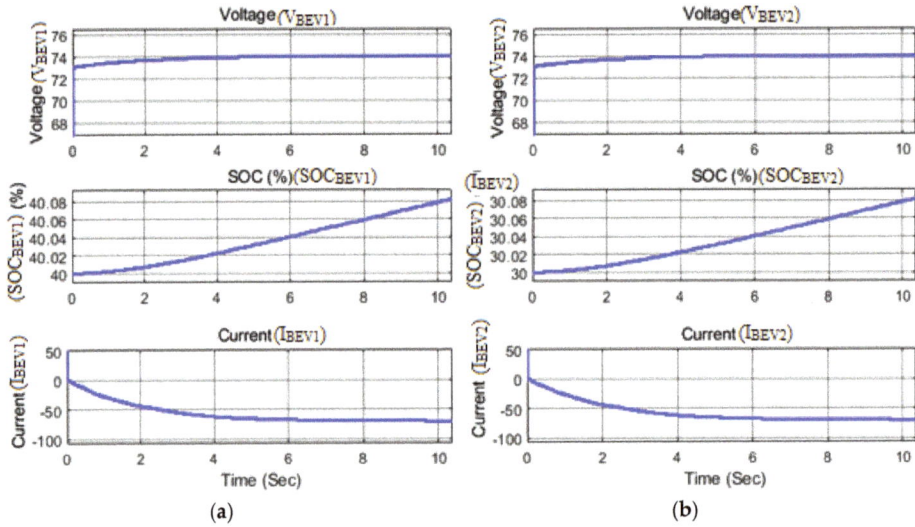

Figure 14. Battery characteristics while charging for electric vehicles (**a**) EV1 and (**b**) EV2.

From the results, it can be observed that the developed power flow optimization algorithm, effectively identifies the availability of power generated by various sources and schedules them as per the load requirement. The developed algorithm achieves fast switching between various sources in order to satisfy the loads continuously without any interruption. Further, the developed algorithm can also be adapted for improving the lifetime of the energy storage device, which is integrated with the system.

7. Conclusions

In this paper, an optimized power flow algorithm for grid-connected photovoltaic system with an energy storage device for charging of electric vehicles is realized. The following objectives are achieved throughout the paper:

1. The need to adopt renewable energy systems for the charging of electric vehicles is studied.
2. The advantage of using an energy storage device with grid integrated PV systems for charging EVs is depicted.
3. The major role of converters in achieving optimal and bidirectional power flow between various sources and loads is realized, and a multiport converter is designed for achieving the objective
4. The need for an intelligent-energy management system for operating the power converter as per the availability of power from the sources and demand on the load side is achieved by developing an optimal power flow algorithm.
5. The developed algorithm is tested with a simulation setup by observing the various modes of operation on the system.

All results and conclusions are based on the assumption that the model created in MATLAB is implemented as a system that adjusts EV charging. A grid-connected PV–EV charging system with an energy storage device is developed in this paper. A multiport converter is designed to accommodate efficient power flow between multiple sources and the loads. The developed converter is controlled by adapting particle swarm optimization, which acts as an intelligent energy management system for dynamic power management at the charging station. An objective function is formulated for scheduling optimal power flow between various sources and loads in the system. The simulation results depict that the new control strategy efficiently shifts between the sources within a stipulated time and continuously supports the charging station.

Author Contributions: Conceptualization, methodology, software, validation, and formal analysis, M.J., V.S.B.K. and M.A.K.; Investigation, A.H. and F.B.; Resources, and data curation, M.J.; writing—original draft preparation, writing—review and editing, and visualization, V.S.B.K. and M.A.K.; supervision, A.H. and F.B.; project administration, A.H. and F.B.

Funding: This research received no external funding.

Conflicts of Interest: The authors declare no conflict of interest.

References

1. Chan, C.C.; Wong, Y.S. Electric Vehicles Charge Forward. *IEEE Power Energy Mag.* **2004**, *2*, 24–33. [CrossRef]
2. Hidrue, M.K.; Parsons, G.R.; Kempton, W.; Gardner, M.P. Willingness to pay for electric vehicles and their attributes. *Resour. Energy Econ.* **2011**, *33*, 686–705. [CrossRef]
3. Yilmaz, M.; Krein, P.T. Review of battery charger topologies, charging power levels, and infrastructure for plug-in electric and hybrid vehicles. *IEEE Trans. Power Electron.* **2013**, *28*, 2151–2169. [CrossRef]
4. Fairley, P. Speed bumps ahead for electric-vehicle charging. *IEEE Spectr.* **2010**, *47*, 13–14. [CrossRef]
5. Clement-Nyns, K.; Haesen, E.; Driesen, J. The impact of Charging plug-in hybrid electric vehicles on a residential distribution grid. *IEEE Trans. Power Syst.* **2010**, *25*, 371–380. [CrossRef]
6. Yunus, K.; la Parra, H.Z.D.; Reza, M. Distribution Grid Impact of Plug-In Electric Vehicles Charging at Fast Charging Stations Using Stochastic Charging Model. In Proceedings of the 2011 14th European Conference on Power Electronics and Applications, Birmingham, UK, 30 August–1 September 2011; pp. 1–11.
7. Botsford, C.; Szczepanek, A. Fast Charging vs. Slow Charging: Pros and cons for the New Age of Electric Vehicles. In Proceedings of the EVS24 International Battery, Hybrid and Fuel Cell Electric Vehicle Symposium, Stavanger, Norway, 13–16 May 2009.
8. Gnann, T.; Funke, S.; Jakobsson, N.; Plotz, P.; Sprei, F.; Bennehag, A. Fast charging infrastructure for electric vehicles: Today's situation and future needs. *Transp. Res. Part D Transp. Environ.* **2018**, *62*, 314–329. [CrossRef]
9. Morrow, K.; Karner, D.; Francfort, J. *Plug-in Hybrid Electric Vehicle Charging Infrastructure Review (U. S. Department of Energy Vehicle Technologies Program—Advanced Vehicle Testing Activity)*; Idaho National Laboratory (INL): Idaho Falls, ID, USA, 2008.
10. Deb, S.; Tammi, K.; Kalita, K.; Mahanta, P. Review of recent trends in charging infrastructure planning for electric vehicles. *Wiley Interdiscip. Rev. Energy Environ.* **2018**, *7*, e306. [CrossRef]
11. Dickerman, L.; Harrison, J. A New Car, a New Grid. *IEEE Power Energy Mag.* **2010**, *8*, 55–61. [CrossRef]
12. Badea, G.; Felseghi, R.A.; Varlam, M.; Filote, C.; Culcer, M.; Iliescu, M.; Răboacă, M. Design and simulation of romanian solar energy charging station for electric vehicles. *Energies* **2019**, *12*, 74. [CrossRef]
13. IEA PVPS. *Trends 2015 in Photovoltaic Applications*; IEA PVPS: Bangkok, Thailand, 2015.
14. Nehrir, M.H.; Wang, C.; Strunz, K.; Aki, H.; Ramakumar, R.; Bing, J.; Miao, Z.; Salameh, Z. A review of hybrid renewable/alternative energy systems for electric power generation: Configurations, control, and applications. *IEEE Trans. Sustain. Energy* **2011**, *2*, 392–403. [CrossRef]
15. Strunz, K.; Abbasi, E.; Huu, D.N. DC microgrid for wind and solar power integration. *IEEE J. Emerg. Sel. Top. Power Electron.* **2014**, *2*, 115–126. [CrossRef]
16. Cingoz, F.; Elrayyah, A.; Sozer, Y. Optimized Resource Management for PV-Fuel-Cell-Based Microgrids Using Load Characterizations. *IEEE Trans. Ind. Appl.* **2016**, *52*, 1723–1735. [CrossRef]
17. Varaha, K.; Bharath, S.; Khan, M.A. *Applications of Artificial Intelligence Techniques in Engineering 698*; Springer: Singapore, 2019.
18. Kurukuru, V.S.B.; Khan, M.A.; Singh, R. Performance optimization of UPFC assisted hybrid power system. In Proceedings of the 2018 IEEMA Engineer Infinite Conference (eTechNxT), New Delhi, India, 13–14 March 2018; pp. 1–6.

19. Eren, Y.; Erdinc, O.; Gorgun, H.; Uzunoglu, M.; Vural, B. A fuzzy logic based supervisory controller for an FC/UC hybrid vehicular power system. *Int. J. Hydrog. Energy* 2009, *34*, 8681–8694. [CrossRef]
20. Badawy, M.O.; Sozer, Y. Power Flow Management of a Grid Tied PV-Battery System for Electric Vehicles Charging. *IEEE Trans. Ind. Appl.* 2017, *53*, 1347–1357. [CrossRef]
21. Carbone, R. Grid-connected photovoltaic systems with energy storage. In Proceedings of the 2009 International Conference on Clean Electrical Power (ICCEP), Capri, Italy, 9–11 June 2009; pp. 760–767.
22. Beltran, H.; Pérez, E.; Aparicio, N.; Rodriguez, P. Daily solar energy estimation for minimizing energy storage requirements in PV power plants. *IEEE Trans. Sustain. Energy* 2013, *4*, 474–481. [CrossRef]
23. F. OECD/IEA. *Medium-Term Market Report/Aut*; IEA: Paris, France, 2014.
24. Deb, S.; Tammi, K.; Kalita, K.; Mahanta, P. Impact of electric vehicle charging station load on distribution network. *Energies* 2018, *11*, 178. [CrossRef]
25. Bauer, P.; Zhou, Y.; Doppler, J.; Stembridge, N. Charging of Electric Vehicles and Impact on the Grid. In Proceedings of the 13th Mechatronika 2010, Trencianske Teplice, Slovakia, 2–4 June 2010; pp. 121–127.
26. SAE. *SAE Electric Vehicle and Plug-in Hybrid Electric Vehicle Conductive Charge Coupler J1772*; SAE: Warrendale, PA, USA; Troy, MI, USA, 2010.
27. IEC 62196. *Plugs, Socket-Outlets, Vehicle Connectors and Vehicle Inlets—Conductive Charging of Electric Vehicles—Part 1, 2, 3*; International Electrotechnical Commission: Geneva, Switzerland, 2014.
28. IEC 61851. *Electric Vehicle Conductive Charging System—Part 1, 21, 23, 24*; International Electrotechnical Commission: Geneva, Switzerland, 2014.
29. Kubel, M. *Design Guide for Combined Charging System*; CharIN e.V: Berlin, Germany, 2015.
30. CHA Association. *Technical Specifications of Quick Charger for the Electric Vehicle*; CHA Association: Washington, DC, USA, 2010.
31. Tesla EV Charging and Supercharging Technique. 2019. Available online: www.teslamotors.com (accessed on 24 April 2019).
32. Messagie, M.; Boureima, F.-S.; Coosemans, T.; Macharis, C.; van Mierlo, J. A Range-Based Vehicle Life Cycle Assessment Incorporating Variability in the Environmental Assessment of Different Vehicle Technologies and Fuels. *Energies* 2014, *7*, 1467–1482. [CrossRef]
33. Messenger, R.A.; Abtahi, A. *Photovoltaic Systems Engineering*; CRC Press: Boca Raton, FL, USA, 2017.
34. Stapleton, G.; Neill, S. *Grid-Connected Solar Electric Systems: The Earthscan Expert Handbook for Planning, Design and Installation*; Routledge: Abingdon, UK, 2011.
35. Carli, G.; Williamson, S.S. Technical Considerations on Power Conversion for Electric and Plug-in Hybrid Electric Vehicle Battery Charging in Photovoltaic Installations. *IEEE Trans. Power Electron.* 2013, *28*, 5784–5792. [CrossRef]
36. Rasin, Z.; Rahman, M.F.; Teknikal, U. Grid-connected Quasi-Z-Source PV Inverter for Electric Vehicle Charging Station. In Proceedings of the International Conference on Renewable Energy Research and Applications, Madrid, Spain, 20–23 October 2013; pp. 627–632.
37. Rasin, Z.; Ahsanullah, K.; Rahman, M.F. Design and Simulation of Quasi-Z-Source Grid-connected PV Inverter with Bidirectional Power Flow for Battery Storage Management. In Proceedings of the IECON 2013-39th Annual Conference of the IEEE Industrial Electronics Society, Vienna, Austria, 10–13 November 2013; pp. 1589–1594.
38. Rydh, C J ; Sandén, B.A. Energy analysis of batteries in photovoltaic systems. Part II: Energy return factors and overall battery efficiencies. *Energy Convers. Manag.* 2005, *46*, 1980–2000. [CrossRef]
39. Chiang, S.J.; Chang, K.T.; Yen, C.Y. Residential Photovoltaic Energy Storage System. *IEEE Trans. Ind. Electron.* 1998, *45*, 385–394. [CrossRef]
40. Riffonneau, Y.; Bacha, S.; Barruel, F.; Ploix, S. Optimal power flow management for grid connected PV systems with batteries. *IEEE Trans. Sustain. Energy* 2011, *2*, 309–320. [CrossRef]
41. Bagula, A.; Castelli, L.; Zennaro, M. On the Design of Smart Parking Networks in the Smart Cities: An Optimal Sensor Placement Model. *Sensors* 2015, *15*, 15443–15467. [CrossRef] [PubMed]
42. Macal, C.M.; North, M.J. Tutorial on Agent-based Modeling and Simulation. In Proceedings of the 37th Winter Simulation Conference, Orlando, FL, USA, 4–7 December 2005; pp. 2–15. [CrossRef]
43. Hutson, C.; Venayagamoorthy, G.K.; Corzine, K.A. Intelligent Scheduling of Hybrid and Electric Vehicle Storage Capacity in a Parking Lot for Profit Maximization in Grid Power Transactions. In Proceedings of the 2008 IEEE Energy 2030 Conference, Atlanta, GA, USA, 17–18 November 2008; pp. 1–8.

44. Kennedy, J. Particle Swarm Optimization. In *Encyclopedia of Machine Learning*; Springer: Berlin, Germany, 2010; pp. 745–817.
45. Valle, Y.; Venayagamoorthy, G.K.; Mohagheghi, S.; Jean-Carlos Hernandez, R.G. Harley, Particle Swarm Optimization: Basic Concepts, Variants and Applications in Power Systems. *IEEE Trans. Evol. Comput.* **2008**, *12*, 171–195. [CrossRef]
46. Mahindra Everito. Available online: http://www.mahindraverito.com/everito/discover-the-new-verito.aspx (accessed on 5 May 2019).
47. Li, G.; Liu, C.C.; Mattson, C.; Lawarrée, J. Day-ahead electricity price forecasting in a grid environment. *IEEE Trans. Power Syst.* **2007**, *22*, 266–274. [CrossRef]
48. Pindoriya, N.M.; Singh, S.N.; Singh, S.K. An adaptive wavelet neural network-based energy price forecasting in electricity markets. *IEEE Trans. Power Syst.* **2008**, *23*, 1423–1432. [CrossRef]
49. Mazadi, M.; Rosehart, W.; Zareipour, H. Impact of Wind Generation on Electricity Markets: A Chance-Constrained Nash Cournot model. *Iran. J. Sci. Technol. Trans. Electr. Eng.* **2012**, *36*, 51–66.
50. Hoke, A.; Brissette, A.; Smith, K.; Pratt, A.; Maksimovic, D. Accounting for lithium-ion battery degradation in electric vehicle charging optimization. *IEEE J. Emerg. Sel. Top. Power Electron.* **2014**, *2*, 691–700. [CrossRef]
51. Motamedi, A.; Zareipour, H.; Rosehart, W.D. Electricity price and demand forecasting in smart grids. *IEEE Trans. Smart Grid* **2012**, *3*, 664–674. [CrossRef]
52. Liu, S.; Wang, J.; Liu, Q.; Tang, J.; Liu, H.; Fang, Z. Deep-Discharging Li-Ion Battery State of Charge Estimation Using a Partial Adaptive Forgetting Factors Least Square Method. *IEEE Access* **2019**, *7*, 47339–47352. [CrossRef]
53. Yang, Y.; Hu, X.; Qing, D.; Chen, F. Arrhenius equation-based cell-health assessment: Application to thermal energy management design of a HEV NiMH battery pack. *Energies* **2013**, *6*, 2709–2725. [CrossRef]
54. Markel, T.; Smith, K.; Pesaran, A.A. *Improving Petroleum Displacement Potential of PHEVs Using Enhanced Charging Scenarios*; Elsevier: Amsterdam, The Netherlands, 2010.
55. Xu, B. *Degradation-Limiting Optimization of Battery Energy Storage Systems Operation*; Location Power Systems Laboratory ETH Zurich: Zürich, Switzerland, 2013.
56. Detailed Model of a 100-kW Grid-Connected PV Array—MATLAB & Simulink—MathWorks France. Available online: https://fr.mathworks.com/help/physmod/sps/examples/detailed-model-of-a-100-kw-grid-connected-pv-array.html (accessed on 5 May 2019).
57. Ma, C.; Marten, F.; Töbermann, J.; Braun, M.; Iwes, F. Evaluation of Modeling and Simulation Complexity on Studying the Impacts of Electrical Vehicles Fleets in Distribution Systems. In Proceedings of the 2014 Power Systems Computation Conference, Wroclaw, Poland, 18–22 August 2014.
58. Mouli, G.R.C. Charging Electric Vehicles from Solar Energy: Power Converter, Charging Algorithm and System Design. Ph.D. Thesis, Delft University of Technology, Delft, The Netherlands, 2018.
59. Mouli, G.R.C.; Bauer, P.; Zeman, M. System design for a solar powered electric vehicle charging station for workplaces. *Appl. Energy* **2016**, *168*, 434–443. [CrossRef]
60. Apostolaki-Iosifidou, E.; Codani, P.; Kempton, W. Measurement of power loss during electric vehicle charging and discharging. *Energy* **2017**, *127*, 730–742. [CrossRef]
61. Wu, T.F.; Chang, C.H.; Chang, Y.D.; Lee, K.Y. Power loss analysis of grid connection photovoltaic systems. In Proceedings of the 2009 International Conference on Power Electronics and Drive Systems (PEDS), Taipei, Taiwan, 2–5 November 2009; pp. 326–331.

© 2019 by the authors. Licensee MDPI, Basel, Switzerland. This article is an open access article distributed under the terms and conditions of the Creative Commons Attribution (CC BY) license (http://creativecommons.org/licenses/by/4.0/).

Article

Integration of Stationary Batteries for Fast Charge EV Charging Stations

Davide De Simone * and Luigi Piegari

Dipartimento di Elettronica, Informazione e Bioingegneria, Politecnico di Milano, 20133 Milano, Italy; luigi.piegari@polimi.it
* Correspondence: davide.desimone@polimi.it

Received: 4 November 2019; Accepted: 4 December 2019; Published: 6 December 2019

Abstract: One of the biggest issues preventing the spread of electric vehicles is the difficulty in supporting distributed fast charging stations by actual distribution grids. Indeed, a significant amount of power is required for fast charging, especially if multiple vehicles must be supplied simultaneously. A possible solution to mitigate this problem is the installation of auxiliary batteries in the charging station to support the grid during high peak power demands. Nevertheless, the integration of high-voltage batteries with significant power is not a trivial task. This paper proposes the configuration and control of a converter to integrate batteries in a fast charging station. The proposed configuration makes it possible to decouple the grid power from the vehicle power using several auxiliary battery modules. At the same time, the converter makes it possible to draw different amounts of power from the battery modules, allowing the use of second life batteries performing in different ways. This paper discusses the design, control, and operation of the converter. Moreover, the effectiveness of the proposed control is shown by means of numerical results.

Keywords: battery electric vehicle; BEV; charging station; stationary storage; plug-in; EV; fast charge; power quality; second life batteries

---

1. Introduction

Electric vehicles (EVs) are spreading in the private transportation sector. Although these vehicles contribute less to city pollution, their spread is being hindered by limitations such as their high price, short range, and long recharge time compared to the equivalent category of internal combustion vehicles [1]. Battery technologies are being rapidly developed, with steady increases in their energy and power densities and decreases in their prices [2].

Higher power density batteries can accept higher recharging power levels, allowing a fast charge. Together with the spread of electric vehicles, the recharging infrastructure is undergoing significant development in terms of charging points and installed power. A generic EV power profile during the recharge phase is characterized by a high peak power demand that usually lasts 26–32 min [3]. Standard fast dc charging is performed at 50 kW dc, but multiple charging systems make it possible to charge at higher power. By 2021, fast charging up to 350 kW might be possible [4]. The most common charging standards and relative power ratings are summarized in Table 1 [4].

In urban areas, it is likely that many consumers will recharge their vehicles between 7:00 and 9:00, at noon, and between 18:00 and 20:00 [3,5]. Grid operators might not be able to overcome future peak power requirements without greatly oversizing the installed power. A possible approach to mitigate this problem might be to add a battery electric storage system (BESS) to a fast charging station (FSS), which will act as a buffer to reduce the peak power demand on the network without increasing the EV charging time.

Table 1. Charging standards.

System	kW	Availability
Combined Charging System (CCS)	50–350	United States, European Union, Australia, Korea
China GB/T	237.5	China, India
Tesla Supercharger	135	Global
CHAdeMO	50–100	Global

Two possible approaches could be followed to interface a BESS with a charging station [6]: ac coupling and dc coupling. In ac coupling, the storage system and charging converters are interconnected through an ac bus. In dc structures, the interconnection is through a common dc link. Figure 1 shows schematics of the two coupling methods.

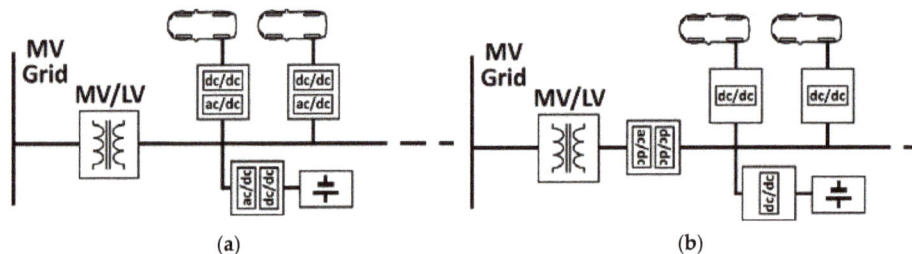

Figure 1. Charging station coupling methods: (a) ac coupling and (b) dc coupling.

At present, ac coupling is the most widespread because of the maturity of the ac technology in terms of the converter design, protection devices, and standards. Although ac coupling is more common, dc coupling provides a higher efficiency and lower cost because this configuration requires fewer conversion stages.

Modular multilevel converters (MMCs) are widely used in dc/ac medium/high-voltage applications. Their multilevel output waveform reduces the total harmonic content of the generated ac voltage and, thus, the harmonics the of ac currents are quite limited also using a very small filter. Depending on the number of levels and the grid impedance, a filter-less solution can be acceptable [7–10].

Double star chopper cells converters (DSCCs) replace the floating capacitors of traditional MMCs with a battery. Recently, the use of this converter in an electric vehicle powertrain has been proposed [11].

Embedding DSCCs in EVs offers some major advantages. The storage system is embedded inside the converter structure, which makes it possible to perform the battery management system (BMS) operations without dedicated hardware [11]. Moreover, it is possible to balance the batteries within each arm to exploit the load currents rather than moving energy from one cell to another, potentially making the process more efficient. The control strategies available in the literature make it possible to recharge the storage system [12] from any power source, i.e., ac three phase, ac single phase, and dc. Finally, the modular structure ensures a highly redundant system [13]. The DSCC structure also seems to be very promising for integrating a storage system in an EV charging station to support the grid during a peak power demand. Indeed, batteries can be integrated, at the module level, directly in the converter modules to ensure the correct power sharing between the modules. In this way, the power converter also operates as a BMS controlling the power exchanged individually by each battery module. Second life batteries can be used because each battery module is controlled separately from the others, and the modularity of the solution makes it possible to substitute only one battery module without the need to substitute all the others. With the proper choice of voltage levels, the dc bus of the DSCC can be used inside the EV station for fast dc charging.

This paper proposes the use of a DSCC converter to integrate an auxiliary storage system with an EV charging station. Moreover, a new control strategy to separately control the power exchanged by this converter at its dc and ac ports is proposed. This control strategy was simulated in a dc coupled charging station, and the results obtained show that is possible to meet the charging power demand by drawing constant power from the three-phase grid. In this way, the power installation required by the EV charging station is limited and lower than the power required by modern fast dc charging systems. This mitigates a big problem with modern fast charging systems, facilitating the diffusion of electric vehicles.

2. Materials and Methods

2.1. Converter Control

The investigated DSCC converter is composed of three phases, each one realized with two arms embedding six modules (Figure 2). The common points of the three phases are used to obtain a virtual dc bus. For each arm, the reference voltage is obtained as follows:

$$\begin{cases} v^*_{lower,k} = \frac{v^*_{dc}}{2} + v^*_{phase,k} + v^*_{circ,k} \\ v^*_{upper,k} = \frac{v^*_{dc}}{2} - v^*_{phase,k} + v^*_{circ,k} \end{cases} \quad (1)$$

where $v^*_{lower,k}$ and $v^*_{upper,k}$ are, respectively, the phase k lower and upper reference arm voltages. v^*_{dc} is the voltage reference for the dc bus of the converter, $v^*_{phase,k}$ is the phase voltage reference, and $v^*_{circ,k}$ is the output of the circulating current controller for phase k used to balance the battery cells of the upper and lower arms [13]. It is worth noting that the three virtual dc buses can have different reference voltages:

$$v^*_{dc,k} = v^*_{dc} + 2 \cdot v^*_{circ,k} \quad (2)$$

This is done to make it possible to balance the state of charge (SOC) of the battery cells of the three legs of the converter.

The modulation strategy must be able to control the dc bus voltage reference. In literature different modulation techniques for this converter topology are available. The modulation techniques can be divided in two categories: nearest level control technique (NLC) and pulse width modulation (PWM) based ones [14]. NLC approximates the output voltage by turning on and off modules in order to minimize the discretization error related to the voltages of the modules, while PWM techniques add a switching signal to the output of NLC. As a consequence, NLC has a discretization error on the dc bus voltage equal to one module voltage. Therefore, if the number of levels is not very high, it can be used neither for a precise control of the dc bus voltage nor for proper balancing among the phases. This is the reason why, in this paper a PWM based technique is used. Given the references for each arm voltage, the control signals of the converter modules are obtained using phase disposition pulse width modulation (PD-PWM) [15]. It is worth noting that the results of this paper can be obtained with any PWM based technique. The choice of the PD-PWM is only based on its simplicity of implementation. This modulation consists of comparing the references with multiple carriers, where the amplitude is the voltage module, translated from the voltages of the other connected modules to avoid overlapping among them, as shown qualitatively in Figure 3.

As the voltages of the battery modules are never perfectly matched, circulating currents might flow among the phases. Given the arm currents, referring to the symbols of Figure 2, the circulating currents ($i_{circ,k}$) and phase currents ($i_{phase,k}$) are defined as follows:

$$\begin{cases} i_{circ,k} = \frac{1}{2}(i_{upper,k} + i_{lower,k}) \\ i_{phase,k} = i_{upper,k} - i_{lower,k} \end{cases} \quad (3)$$

where $i_{upper,k}$ and $i_{lower,k}$ are the phase k arm currents. It is worth noting that the dc component of the circulating current is used to balance the batteries among the phases, while energy is exchanged between the upper and lower arms using the second harmonic component of the same circulating current. A dedicated circulating current controller is implemented to allow battery balancing among the arms and phases [13]. The leg and arm balance is achieved following [16]. The logic scheme of the circulating current controller is shown in Figure 4, where v_{ph} is the space vector of the phase voltages generated by the converter; i^*_{circ} is the reference space vector of the circulating currents; and $v^*_{circ,a}$, $v^*_{circ,b}$, and $v^*_{circ,c}$ are the dc bus references for each phase. In practice, the balancing is obtained with three different actions:

1. Balancing among battery modules belonging to same arm is achieved by sorting the cells on the basis of their SOC and using the most charged (discharged) when the current is discharging (charging) them;
2. Balancing among battery modules belonging to different legs is achieved by changing the dc reference voltage of each leg allowing the circulation of a dc current among the legs (this current does not interest the load);
3. Balancing among upper and lower arms is achieved using an oscillating zero-sequence current (equal on each phase) that takes more energy from the lower (upper) arm if this is the more charged.

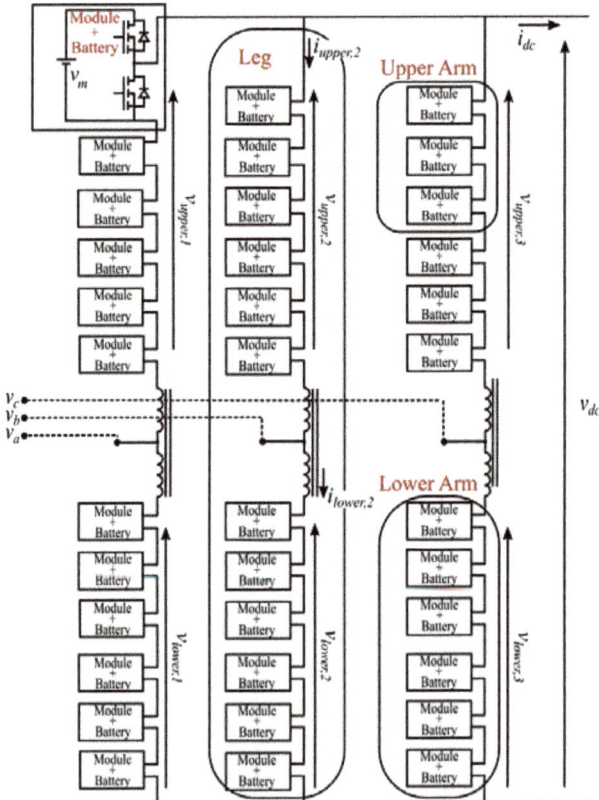

Figure 2. Six modules per arm of the double star chopper cells converter (DSCC).

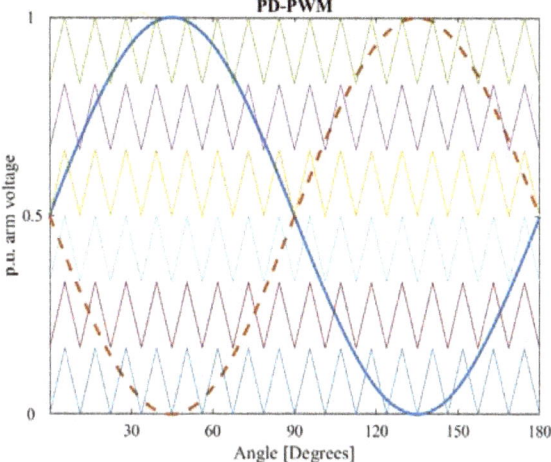

Figure 3. Phase disposition PWM (PD-PWM) qualitative modulation. Solid line: upper arm reference, dashed line: lower arm reference.

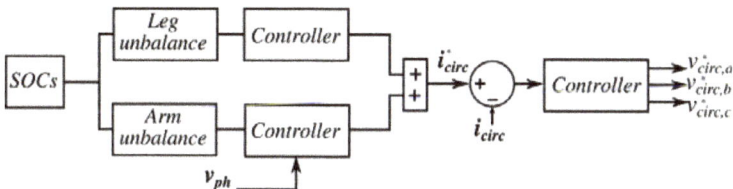

Figure 4. Leg and arm balance controllers. SOCs are all the modules' state of charge values. "Leg unbalance" computes the mean SOC of each phase and compares the results. "Arm unbalance" computes the imbalance between the upper and lower arms for each phase.

This balancing technique, anyway, is widely addressed in [16] and, therefore, it will not analyzed in more details in this paper.

In order to control the dc power output of the converter, a dedicated controller acts on the average dc bus voltage of the converter: reference dc current i^*_{dc} is computed given reference dc power P^*_{dc} and dc bus voltage v_{dc}. The difference between the reference and actual dc current, i_{dc}, is used for the input of the dc current regulator acting on the converter virtual dc bus voltage reference v^*_{dc}, as reported in Figure 5.

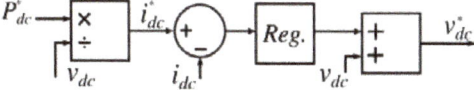

Figure 5. dc power regulator.

The ac power is controlled on a reference frame oriented on the grid voltage by means of a phase locked loop (PLL). On this reference frame, the controls of the active and reactive power are decoupled. Indeed, the direct axis oriented on the voltage space vector is indicated with "d" and the axis with a 90° lead is shown by "q", where the quadrature component of the voltage is equal to zero, resulting in the following [17,18]:

$$P_{ac} = \tfrac{3}{2} v_d i_d$$
$$Q_{ac} = \tfrac{3}{2} v_d i_q$$
(4)

Therefore, the control of the active power is obtained by dividing the active power reference P^*_{ac} by the direct component of the grid voltage vector, obtaining the direct current reference i^*_d. Using a multilevel converter, the filter impedance to connect to the ac grid is very low, and it is possible to control the direct and quadrature components of the current by directly controlling, respectively, the direct and quadrature components of the voltage produced by the converter.

The position of the voltage grid, $\angle v_{ph}$, is then used to obtain the reference for the space vector of the voltage to be used for the converter. The simplified control scheme is reported in Figure 6.

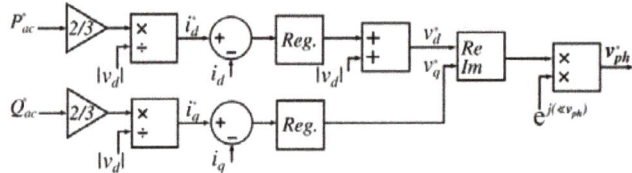

Figure 6. Active and reactive power regulator.

In [3] the charging station exchanges reactive power with the grid to provide grid services and increasing the profitability of the installation. If no requirements is given for the reactive power, its reference is usually set to zero to exchange power at unitary power factor minimizing the current and, consequently, the losses. A zero reference for the reactive power is assumed for this paper.

2.2. Converter Sizing

Acting separately on the ac and dc power references, it is possible to simultaneously control the dc and ac active power levels. The difference between the two is automatically stored or supplied by the storage system. To simultaneously control the two power levels, the voltage levels of the converter must be chosen so that the two controllers do not interact with each other. In particular, the lowest voltage on the dc bus to regulate the power flows must be higher than the EV battery voltages and, at the same time, it has to be higher than twice the highest voltage that is required on the ac side.

Allowing a voltage variation on the ac side of ±10% and considering an additional ±10% margin to allow the control of the reactive power, the ac output voltage of the converter has to be able to reach a value that is 20% higher than the rated ac voltage. As a consequence, to allow the full controllability on the ac side, the minimum dc voltage $v_{dc,min}$ is as follows:

$$v_{dc,min} = 2 \cdot 1.2 \cdot v_{ph,rated} \tag{5}$$

where $v_{ph,rated}$ is the peak of the rated grid voltage. If a ±10% margin is considered for the regulation of the dc power and the balancing among the phases, the minimum module voltage $v_{module,min}$ can be calculated as follows:

$$v_{module,min} = \frac{1.1 v_{dc,min}}{n_{modules}} \tag{6}$$

where $n_{modules}$ is the number of modules per arm. In [3], a charging station with a storage system was dimensioned so that the batteries' SOC did not go below 20% under normal working conditions. The minimum number of cells in the stack is thus obtained by applying the following:

$$n_{stack} = \left\lceil \frac{v_{module,min}}{v_{cell,min}} \right\rceil \tag{7}$$

where $v_{cell,min}$ is the minimum voltage achieved by one battery cell at the lowest SOC limit.

In this article, three vehicles of different categories are considered: the BMW i3 (EV_1), belonging to the city car category; the Nissan Leaf e+ (EV_2), representing a medium size vehicle; and finally, the

Porsche Taycan (EV3), which is a sports car. Their storage systems and charging power specifications are summarized in Table 2 [4,19–21].

Table 2. Considered vehicles' specifications.

	Vehicle	Usable Battery Capacity (kWh)	Charging Power (kW)
EV_1	BMW i3	37.9	46
EV_2	Nissan Leaf e+	56	100
EV_3	Porsche Taycan	83.7	350

The charging station is supposed to have one fast charging dc port for the vehicles. It draws power from a 100 kW three-phase low-voltage grid. Each vehicle is supposed to arrive at the charging spot with a 10% SOC. The end of charge is considered at a SOC of 90%. As shown in [3], the average time, t_{swap} necessary to connect the charging station to the next customer is 10 min. To grant the maximum charging power to the vehicles, the stationary storage system is dimensioned to be able to recharge the medium size and large EVs in a row. The energy that must be stored is given by the energy balance between the network and the vehicles:

$$E_{storage} = \frac{1}{0.6} \cdot \frac{1}{\eta} \left\{ 0.8 \cdot (E_{EV,2} + E_{EV,3}) - P_{network} \cdot \left[0.8 \cdot \left(\frac{E_{EV,2}}{P_{EV,2}} + \frac{E_{EV,3}}{P_{EV,3}} \right) + \frac{t_{swap}}{60} \right] \right\} \quad (8)$$

where 0.6 represents the exploited energy of the stationary storage (20%–80%); η is the global efficiency of the storage and converter, which is supposed to be 90%; $E_{EV,k}$ and $P_{EV,k}$ are, respectively, the battery rated energy in Wh and the charging power in W of the k vehicle, while t_{swap} is expressed in minutes.

The power exchange with the grid is managed in order to keep the stationary storage system SOC below 80% and to not overcome the power limit of 100 kW. If the vehicle is being charged at less than 100 kW and the stationary storage reaches 80%, the network power is reduced to match the vehicle power demand. The dc power controller instead is programmed to prevent the storage system from reaching a SOC lower than 20%. If a vehicle whose charging power is higher than the grid power is connected, the dc power will be the maximum allowed until the storage system reaches a minimum SOC (SOC_{min}) of 20%; then the dc power is limited to the power from the grid. To generate the dc power reference, the stationary battery SOC_{min} is compared with the minimum allowed SOC of 20%. The result is given as an input to a regulator whose output is saturated between the minimum and maximum dc power levels of the installed DSCC. The result is that the maximum dc power $P_{dc,max}$ is compatible with the current SOC_{min}. The dc power required by the vehicle is then saturated between 0 and $P_{dc,max}$. The schematic of the minimum SOC regulator is reported in Figure 7.

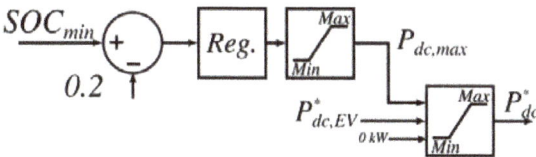

Figure 7. Min SOC regulator.

The ac power reference is generated by comparing the average SOC with the maximum allowed SOC of 80%. The result is given to a regulator whose output is saturated between 0 and the maximum ac power, i.e., 100 kW (Figure 8).

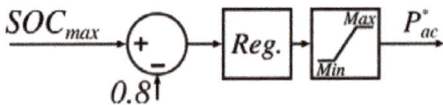

Figure 8. Max SOC regulator.

2.3. Test Setup

The proposed control strategy was tested by simulating the continuous connection of the vehicles one after the other with pauses of 10 min between subsequent connections (as shown in Figure 9b).

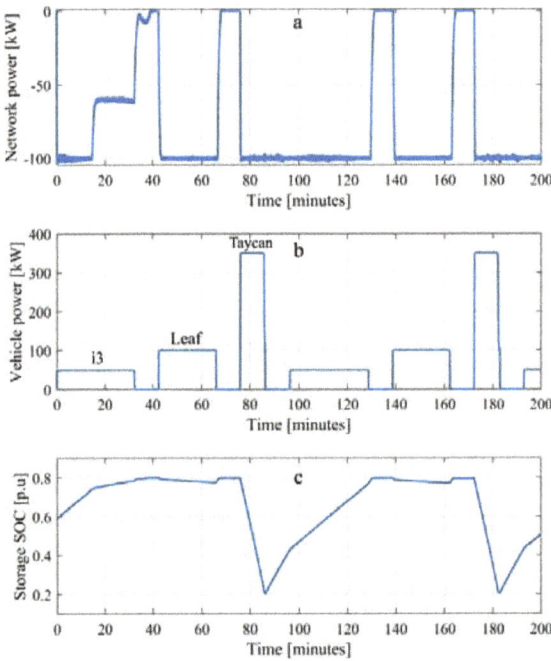

Figure 9. Simulation results: (a) grid power, (b) vehicle power, and (c) average SOC of stationary battery.

The initial average SOC of the charging station battery was considered to be 60%. To test the BMS capabilities, all of the modules were initialized with random SOCs between 40% and 80%.

3. Results

The control strategy was validated using Simulink. Figure 9 shows that the ac and dc power levels could be controlled separately by exploiting the DSCC storage system as an energy buffer, recharging the vehicles at their full capabilities without overloading the grid.

Comparing Table 3 with Table 2, it is possible to state that the full charging power capabilities were achieved even though the stationary storage system had a lower capacity than the highest-range vehicle. Moreover, the stationary storage was sized to maximize its expected lifespan using only an SOC range of 20%–80%.

In Figure 10, it is possible to analyze the effectiveness of the module balancing algorithm. Even if the starting condition represents a very unbalanced situation, the BMS capabilities of the DSCC topology allows the cells to be balanced without dedicated hardware in the first 40 min. From the analysis of the figure, it is clear that, in the first 5 min, the balancing inside each arm is completed by using the load current.

Table 3. Summary of the parameters used for the simulations.

Converter Parameters		
Parameter	Value	Meas. Unit
$v_{ph,rated}$	311	V
$v_{cell,min}$	3	V
$v_{dc,min}$	828	V
$n_{modules}$	6	
n_{stack}	46	
$v_{module,min}$	138	V
$E_{storage}$	69	kWh

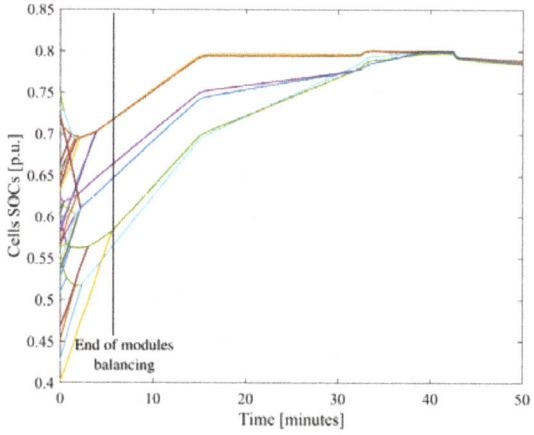

Figure 10. Modules' SOCs.

Then, in a longer time of up to 40 min, the balancing among the arms is achieved. As the arm balancing relies on the injection of circulating currents, the balancing power can be set to maximize the efficiency or reduce the balancing time. It is worth noting that SOCs of all the modules is always kept under the limit of 80% following the control rule presented in Section 3.

4. Discussion

At present, the main obstacles to the diffusion of EVs are their high costs, limited vehicle ranges, and long recharging times. In the future, several efforts will be devoted to increase the vehicle range while reducing the recharging time. For this reason, increasingly higher power levels will be required for charging stations. Nevertheless, the widespread requests for high-power nodes represent a critical scenario for actual electric transmission and distribution grids. For this reason, the option to add a local storage system in proximity to a fast charging station has been under analysis in the recent literature. In this paper, a new configuration to integrate battery modules in a fast charging station was proposed and studied. In particular, a DSCC power converter was used to achieve the following goals: (i) increasing the dc output voltage using lower voltage battery modules and (ii) balancing the battery modules by separately controlling the power drawn by each module, allowing the use of second life batteries. Moreover, a control strategy for the power converter was proposed in order to decouple the control of the ac and dc power values exchanged by the converter. With the proposed control, it is possible to exchange two different power values on the two ports of the converter, and the power difference is automatically supplied or drawn by the stationary battery. In this way, it is possible to increase the power available for the fast charging station, requiring only the average power from the grid.

In this study, a power converter was modeled, and numerical results were obtained for a single point fast charging station. A sequence of three different kinds of vehicles, one city car, one medium-size car, and one sports car, were simulated, and the obtained numerical results showed the effectiveness of the proposed converter and strategy. Moreover, the balancing capabilities of the power converter were demonstrated.

Author Contributions: Conceptualization, D.D.S.; methodology, D.D.S. and L.P.; software, D.D.S.; validation, D.D.S. and L.P.; formal analysis, D.D.S.; investigation, D.D.S. and L.P.; resources, D.D.S. and L.P.; data curation, D.D.S.; writing—original draft preparation, D.D.S.; writing—review and editing, L.P.; visualization, D.D.S.; supervision, L.P.; project administration, L.P.; funding acquisition, L.P.

Funding: This research received no external funding.

Conflicts of Interest: The authors declare no conflict of interest.

References

1. Hoimoja, H.; Rufer, A.; Dziechciaruk, G.; Vezzini, A. An ultrafast EV charging station demonstrator. In Proceedings of the International Symposium on Power Electronics Power Electronics, Electrical Drives, Automation and Motion, Sorrento, Italy, 20–22 June 2012; pp. 1390–1395.
2. Overview of Battery Cell Technologies, Marcel MEEUS. Energy Materials Industrial Research Initiative (EMIRI). Available online: https://europa.eu/sinapse/webservices/dsp_export_attachement.cfm?CMTY_ID=0C46BEEC-C689-9F80-54C7DD45358D29FB&OBJECT_ID=230DABFD-90AB-8F7D-083EF5BD909DD025&DOC_ID=9C5B91FE-01BC-5F72-79D01E1939A9EE53&type=CMTY_CAL (accessed on 8 October 2019).
3. Richard, L.; Petit, M. Fast charging station with battery storage system for EV: Grid services and battery degradation. In Proceedings of the 2018 IEEE International Energy Conference (ENERGYCON), Limassol, Cyprus, 3–7 June 2018; pp. 1–6.
4. Voelcker, J. Porsche's fast-charge power play: The new, all-electric Taycan will come with a mighty thirst. This charging technology will slake it. *IEEE Spectr.* **2019**, *56*, 30–37. [CrossRef]
5. Jochem, P.; Landes, P.; Reuter-Oppermann, M.; Fichtner, W. Workload Patterns of Fast Charging Stations Along the German Autobahn. *World Electr. Veh. J.* **2016**, *8*, 936–942. [CrossRef]
6. Srdic, S.; Lukic, S. Toward Extreme Fast Charging: Challenges and Opportunities in Directly Connecting to Medium-Voltage Line. *IEEE Electrif. Mag.* **2019**, *7*, 22–31. [CrossRef]
7. Martinez-Rodrigo, F.; Ramirez, D.; Rey-Boue A., B.; de Pablo, S.; Herrero-de Lucas L., C. Modular Multilevel Converters: Control and Applications. *Energies* **2017**, *10*, 1709. [CrossRef]
8. Van Hertem, D.; Gomis-Bellmunt, O.; Liang, J. *HVDC Grids: For Offshore and Supergrid of the Future*; John Wiley & Sons: New York, NY, USA, 2016.
9. Li, J.; Bhattacharya, S.; Huang, A.Q. A New Nine-Level Active NPC (ANPC) Converter for Grid Connection of Large Wind Turbines for Distributed Generation. *IEEE Trans. Power Electron.* **2011**, *26*, 961–972. [CrossRef]
10. Ruderman, A.; Reznikov, B. Time domain evaluation of filterless grid-connected multilevel PWM converter voltage quality. In Proceedings of the 2010 IEEE International Symposium on Industrial Electronics, Bari, Italy, 4–7 July 2010; pp. 2940–2945.
11. D'Arco, S.; Piegari, L.; Tricoli, P. A modular converter with embedded battery cell balancing for electric vehicles. In Proceedings of the 2012 Electrical Systems for Aircraft, Railway and Ship Propulsion, Bologna, Italy, 16–18 October 2012; pp. 1–6.
12. D'Arco, S.; Quraan, M.; Tricoli, P.; Piegari, L. Battery charging for electric vehicles with modular multilevel traction drives. In Proceedings of the 7th IET International Conference on Power Electronics, Machines and Drives (PEMD 2014), Manchester, UK, 8–10 April 2014; p. 2.
13. Quraan, M.; Yeo, T.; Tricoli, P. Design and Control of Modular Multilevel Converters for Battery Electric Vehicles. *IEEE Trans. Power Electron.* **2015**, *31*, 507–517. [CrossRef]

14. De Simone, D.; Piegari, L.; D'Areo, S. Comparative Analysis of Modulation Techniques for Modular Multilevel Converters in Traction Drives. In Proceedings of the 2018 International Symposium on Power Electronics, Electrical Drives, Automation and Motion (SPEEDAM), Amalfi, Italy, 20–22 June 2018; pp. 593–600.
15. Ouerdani, I.; Bennani, A.; Ben, A.; Slama, B.; Montesinos Miracle, D. Phase Opposition Disposition PWM Strategy and Capacitor Voltage Control for Modular Multilevel Converters. In Proceedings of the International Conference on Recent Advances in Electrical Systems, Hammamet, Tunisia, 20–22 December 2016.
16. Brando, G.; Dannier, A.; Spina, I.; Tricoli, P. Integrated BMS-MMC Balancing Technique Highlighted by a Novel Space-Vector Based Approach for BEVs Application. *Energies* **2017**, *10*, 1628. [CrossRef]
17. Zahoor, W.; Zaidi, S.H. Synchronization and dq current control of grid-connected voltage source inverter. In Proceedings of the 17th IEEE International Multi Topic Conference 2014, Karachi, Pakistan, 8–10 December 2014; pp. 462–466.
18. Piegari, L.; Tricoli, P. A control algorithm of power converters in smart-grids for providing uninterruptible ancillary services. In Proceedings of the 14th International Conference on Harmonics and Quality of Power - ICHQP 2010, Bergamo, Italy, 26–29 September 2010; pp. 1–7.
19. Electric Vehicle Database. Available online: https://ev-database.org/car/1145/BMW-i3-120-Ah (accessed on 30 September 2019).
20. Electric Vehicle Database. Available online: https://ev-database.org/car/1144/Nissan-Leaf-eplus (accessed on 30 September 2019).
21. Electric Vehicle Database. Available online: https://ev-database.org/car/1116/Porsche-Taycan-Turbo-S (accessed on 30 September 2019).

© 2019 by the authors. Licensee MDPI, Basel, Switzerland. This article is an open access article distributed under the terms and conditions of the Creative Commons Attribution (CC BY) license (http://creativecommons.org/licenses/by/4.0/).

Article

Location and Sizing of Battery Energy Storage Units in Low Voltage Distribution Networks

Andrea Mazza [1], Hamidreza Mirtaheri [2], Gianfranco Chicco [1,*], Angela Russo [1] and Maurizio Fantino [2]

1. Dipartimento Energia "Galileo Ferraris", Politecnico di Torino, 10129 Turin, Italy; andrea.mazza@polito.it (A.M.); angela.russo@polito.it (A.R.)
2. Links Foundation, 10138 Torino, Italy; hamidreza.mirtaheri@linksfoundation.com (H.M.); maurizio.fantino@linksfoundation.com (M.F.)
* Correspondence: gianfranco.chicco@polito.it

Received: 12 November 2019; Accepted: 17 December 2019; Published: 20 December 2019

Abstract: Proper planning of the installation of Battery Energy Storage Systems (BESSs) in distribution networks is needed to maximize the overall technical and economic benefits. The limited lifetime and relatively high cost of BESSs require appropriate decisions on their installation and deployment, in order to make the best investment. This paper proposes a comprehensive method to fully support the BESS location and sizing in a low-voltage (LV) network, taking into account the characteristics of the local generation and demand connected at the network nodes, and the time-variable generation and demand patterns. The proposed procedure aims to improve the overall network conditions, by considering both technical and economic aspects. An original approach is presented to consider both the planning and scheduling of BESSs in an LV system. This approach combines the properties of metaheuristics for BESS sizing and placement with a greedy algorithm to find viable BESS scheduling in a relatively short time considering a specified time horizon, and the application of decision theory concepts to obtain the final solution. The decision theory considers various scenarios with variable energy prices, the diffusion of local renewable generation, evolution of the local demand with the integration of electric vehicles, and a number of planning alternatives selected as the solutions with top-ranked objective functions of the operational schedules in the given scenarios. The proposed approach can be applied to energy communities where the local system operator only manages the portion of the electrical grid of the community and is responsible for providing secure and affordable electricity to its consumers.

Keywords: distribution system; batteries; storage; planning; scheduling; decision theory

1. Introduction

The progress of technologies concerning different types of batteries and their control systems, together with the evolution of a regulatory framework in which energy storage is considered more explicitly, are making Battery Energy Storage Systems (BESSs) progressively more cost-effective for energy system applications. A BESS is specified by its power rating and energy capacity. Both of these specifications impact the BESS investment cost and need to be defined separately [1]. Typical BESS applications for power and energy systems include improvement of the quality of service, assistance with primary and secondary frequency control to enhance network stability, the smoothing of power fluctuations in the generation profiles for better integration of Renewable Energy Sources (RES), and the promotion of higher users' participation in demand management through time-shifting of the energy usage [2]. Most BESS applications in electrical networks refer to times of 1–2 h and to BESS sizes smaller than 50 kW. However, the number of applications for longer times (e.g., 2–5 h) and sizes up to 500 kW is already significant, and applications for bigger sizes (also in the range 1–10 MW) are increasing.

In the transmission system, the positive impacts of BESSs for providing a fast response to frequency deviations [3], mitigating under-frequency transients [4], and being exploited for energy arbitrage [5] have been observed. The integration of BESSs with appropriate control strategies has been shown to be able to improve the frequency stability [6]. At the distribution system level, a BESS is mainly used for enhancing the grid integration of RES by mitigating the effects of the uncertainty on load and distributed generation [7,8], improving the distribution system reliability by avoiding operations close to the line thermal limits and thus more exposed to the risk of protection trips [9], enhancing the quality of the supply with relatively high-power and low-energy solutions [10], reducing the need for grid expansion, shaving the power demand peaks through load shifting or load leveling [9], optimizing the energy transaction costs [11], and integrating BESSs with the demand response in microgrid applications [12]. In [13], a distinction is made between a centralized BESS (that participates in reducing the demand deviation, avoiding the reverse power flow and correcting the power factor), and a distributed BESS (that aims at annual energy loss reduction, reduction of the demand deviation, and improvement of the voltage profile).

A key challenge is to determine the power rating, energy capacity, and location of BESSs in the distribution network. A number of contributions have been published on the siting and sizing of BESSs. Different objectives, such as peak shaving, voltage regulation, and reduction of the energy not supplied, are combined in [14] in an optimal power flow-based approach. In [15], reliability improvement, together with peak shaving, is considered as an objective. The review [16] presents a categorization of the methods used to determine the BESS siting and sizing, in which four main groups are identified (analytical methods, exhaustive search, mathematical programming, and heuristic methods). Decision-making tools are also applied. For example, in [17], the optimal BESS sizing and siting is identified in a microgrid with RES taking into account demand and generation uncertainty by using decision theory criteria. In [18], hybrid energy storage systems, including BESSs, are addressed, by determining the energy storage capacity in distribution systems through an assessment of the risk tolerance of the investors. The results of BESS installations in terms of providing different services at different voltage levels are reported in [19].

A specific aspect generally not highlighted in the reviews on BESS siting and sizing is the distinction between applications for Medium Voltage (MV) and Low Voltage (LV) systems. The literature contributions mainly refer to Medium Voltage (MV) distribution systems. However, the formulation of a planning problem referring to the installation of BESSs in LV systems has various differences with respect to what happens in MV systems. First of all, in MV systems, the study can be conducted by assigning growth in the local generation and demand aggregated at the MV node level. In this case, it is possible to mix up the contributions from different energy sources at the LV level, without looking at the details of the individual sources. Additionally, the uncertainty that characterizes the local generation is seen with respect to the aggregation of the generations, typically taking into account possible correlations between the generation patterns due to external variables (e.g., solar irradiance and temperature for photovoltaic systems). Then, it is possible to exploit the smoothing effect due to the aggregation of a number of individual demands, define typical patterns for the aggregated demand, and associate these typical patterns to predefined evolutions in time.

Conversely, for an LV system, the level of aggregation of demand and local generation is much lower, and many more critical aspects appear. In particular, the local characteristics of the generation and demand at each LV node have to be considered individually. The uncertainties of local generation and demand increase as the smoothing effect of the aggregations is reduced. The setting up of scenarios of development of the local generation cannot proceed with a generic aggregated effect at each LV node, but has to take into account where there are different types of local generation and what reasonable increment can be established for that type of generation. Some relevant differences between MV and LV systems are summarized in Table 1.

Table 1. Summary of relevant differences between Medium Voltage (MV) and Low Voltage (LV) systems.

Characteristics	MV Grid	LV Grid	Consequences
Structure	Weakly meshed	Radial	LV grid cannot be reconfigured, thus the proper network operation has to be guaranteed thanks to the devices connected to the grid
Load	Balanced	Unbalanced	In the LV grid, it is advisable to apply three-phase load flow
Branch	$R \approx X$	$R >> X$	In the LV grid, the voltage drop is strictly correlated to the active power, and voltage control can thus be effectively conducted by modifying the net nodal active power
Load profile	Aggregate	Not aggregate	LV grid represents the last mile of the grid and supplies the customer. A more detailed representation of loads and generation profiles is then necessary and the aggregation impact is less evident than in MV grids

To achieve the best BESS performance and to maximize the overall benefits, proper planning is necessary. For instance, in [20], several configurations of BESSs are compared and the overall network impact is evaluated and compared for different placements of BESSs in the network. In the relevant literature, there are some contributions related to the optimal planning (sizing and placement) of energy storage systems in LV distribution networks. The method applied in [21] aims to optimally configure the energy storage systems to alleviate over- and under-voltage problems. The problem of the optimal location is solved by a heuristic method based on voltage sensitivity analysis. Uncertainties due to stochastic generation and demand are also considered in the optimal sizing and the worst-case approach is applied to select the sizes.

Some contributions refer to LV distribution networks characterized by a high penetration of photovoltaic generation, and consider the possibility of alleviating the negative impacts by the installation of storage systems. In [22], a heuristic method is applied to determine the optimal location and sizing of storage systems and the objective to be minimized is a cost function accounting for the cost of storage systems and the cost due to voltage deviations. The optimal planning of BESSs proposed in [23] aims at maximizing an objective function that includes both benefits and costs (i.e., energy arbitrage, environmental emission, energy losses, transmission access fee, capital, and maintenance costs of a BESS). Daily charge/discharge of the storage systems is also determined considering a proper model of the BESS operation. In [24], the authors propose a procedure for the optimal placement and sizing of distributed energy storage systems in low voltage distribution systems aimed at maximizing the utilization of photovoltaic plants and minimizing the battery degradation. The multi-objective optimization problem proposed by [25] is focused, from one side, on the minimization of energy losses and, from the other side, on the minimization of costs associated with distributed generators and energy storage systems. A distribution system with a high penetration of photovoltaics generators is considered in [26]. A heuristic procedure for reducing the search space for the location of storage systems in a low voltage microgrid is proposed in [27], where analytical considerations on the voltage sensitivity, voltage unbalances, and line loading drive are included in the selection of candidate locations.

This paper proposes an overall procedure to address the BESS location and sizing in an LV network, taking into account the characteristics of the local generation and demand connected at the LV nodes and the time-variable generation and demand patterns. The proposed procedure aims to improve the overall network conditions by considering both technical and economic aspects. This condition aims to represent the conditions that could be found in the case of energy communities where the local system operator only manages the electrical grid of the community and is responsible for guaranteeing secure and affordable electricity for its consumers.

An original algorithm is presented to consider both the planning and scheduling of BESSs in an LV system. This algorithm combines the properties of metaheuristics (to explore the solution space as much as possible), and a greedy algorithm able to find viable BESS scheduling in a relatively short period of time, considering a specified time horizon. The final solution is obtained by using a multi-criteria decision making (MCDM) approach based on the application of decision theory concepts [28] to a number of selected planning alternatives evaluated for different weighted scenarios. Decision theory is an appropriate tool for dealing with cases in which the uncertainty on possible future situations is very large and is handled through scenario analysis.

The next sections of this paper are organized as follows. Section 2 describes the details of the methodology used to address the planning problem. Section 3 presents the application to an LV distribution system and discusses the results. The last section contains the concluding remarks.

2. Description of the Methodology

The BESS location and sizing is analysed as a planning problem seen from the point of view of the electricity manager of an energy community, which is responsible for guaranteeing secure and affordable electricity for its customers at a minimum cost, as well as for both the infrastructure and quality of the service.

2.1. Data Resolution and Reference Period

The data used are assumed to have a constant time resolution Δt. A reference period of duration T_{ref} is assumed, in which the operation of the LV distribution system is analysed in detail by considering the generation and demand patterns and a specific model for BESS scheduling. The planning problem is set up for a time horizon multiple of T_{ref}, namely, with an overall duration of $T_H = N_H \times T_{ref}$, where $N_H > 1$ is an integer number. It is assumed that the BESSs that will be chosen by the proposed procedure will be installed at the beginning of the time period of analysis. The planning time horizon chosen is longer than the lifetime of the BESS, in order to include the replacement of the BESS during the planning period.

2.2. Definition of the Scenarios

The electricity prices, diffusion of the local generation, and diffusion of electric vehicles have been assumed to be uncertain data inputs that affect the solution of the planning problem. Several methods can be applied to handle uncertain variables; in this proposal, several scenarios will be identified to represent different instances.

Starting from the results determined for the reference period, a number of scenarios are constructed to represent possible paths of evolution of selected quantities that affect the LV network operation. The scenarios are defined by taking into account the long-term changes in time that may appear in the following quantities:

(a) *Electricity prices*: the consumers or prosumers connected to the LV system are considered as price takers, namely, they do not participate in the definition of energy prices in the wholesale electricity market. M_P trends of variation of the electricity prices are established by the user by considering the final increase (or decrease) of the electricity price at the end of the planning time horizon;

(b) *Diffusion of the local generation*: the distributed generation (DG) connected to the LV network can change at selected locations in different ways. For LV systems, it is likely that more prosumers will install their local generation systems at locations where there is no local production. From the point of view of the scenario definition, M_{DG} trends of variation of the local generation are considered, and each one is defined by assuming a rate of increase of the energy production from local generation (not of the power installed);

(c) *Diffusion of electric vehicles*: in the present situation, the diffusion of EVs is still relatively limited in many jurisdictions. The number of EVs will increase in the future, and different hypotheses about their number can be represented by M_{EV} trends.

Under the hypotheses provided for the scenarios, the number of scenarios considered is equal to $M = M_P \times M_{DG} \times M_{EV}$.

The scenarios obtained are applied to calculate the objective function for the planning problem (Section 2.4), and are weighted in order to be used in the decision-based approach illustrated in Section 2.6.

2.3. Definition of the Sizing Alternatives

Due to the particular use of the BESS, which includes load leveling, the indication of [29] to use the energy to power ratio equal to 2 is followed. In this way, data are described in terms of their energy capacity, and the power rating is then directly linked to the energy capacity through the energy to power ratio.

In the distribution network, there are K nodes, but it is assumed that a user-defined number $K_{BESS} < K$ of nodes is taken into account for possible BESS location. A maximum value for the BESS energy capacity $C_{k,max}$ is assigned to each node $k = 1, \ldots, K_{BESS}$, depending on the characteristics of the node. The final BESS energy capacity to be assigned to each node will be determined by the proposed approach in the range from zero to $C_{k,max}$ at the nodes $k = 1, \ldots, K_{BESS}$. To avoid the use of continuous variables, this range is partitioned into a given number Λ of BESS energy capacity levels. Without loss of generality, the number Λ is chosen as a constant for all the nodes, that is, it is independent of k. In this way, the total number of alternative combinations of BESS sizes is $S = \Lambda^{K_{BESS}}$. Even for relatively small numbers of nodes and BESS energy capacity levels, the number of combinations S can become so high that even their enumeration becomes intractable with an exhaustive search process. For example, if $K_{BESS} = 20$ nodes and $\Lambda = 5$ BESS energy capacity levels, the result is $S = 5^{20} = 9.54 \times 10^{13}$. In this situation, only parts of the combinations will be reached during the planning procedure, for instance, that conducted by using a metaheuristic algorithm able to explore the solution space with a conceptual direction of evolution towards the global optimum of the objective function employed in the definition of the planning problem.

2.4. Definition of the Objective Functions

The solution of the power flow at each time step $h = 1, \ldots, H$, together with the BESS operational schedules, provide the data required to calculate the power flows and the LV network losses, and to determine whether there is a reverse power flow with the power injected at the supply point. These results are used as the contribution of the distribution system operation to the formulation of the objective function for the planning problem.

The general objective function formulated for the planning problem is a penalized objective function defined on the basis of the investment and operation and maintenance (O&M) costs, and of penalty terms associated with violations of the voltage limits and with the presence of reverse power flow to the MV distribution system. The expression of the objective function is constructed, starting from a reference function f_{ref} and including some penalty terms to obtain the penalized objective function f_P:

$$f_{ref} = \Delta c_{inv} + c_{O\&M} \tag{1}$$

$$f_P = f_{ref} \times (1 + \pi_V + \pi_R), \tag{2}$$

where the addends have the following meaning:

- Δc_{inv}: investment costs for BESS purchasing and installation, determined by using the cost per kWh, depending on the BESS energy capacity, and the cost per kW applied to the inverter for grid connection of the BESS (depending on the BESS power rating) [29];

- $c_{O\&M}$: operation and maintenance costs, calculated by considering the costs of network losses (evaluated using the electricity prices), the BESS aging applied as a reduction of the maximum BESS energy capacity [29], and the maintenance costs considered as a percent of the investment costs;
- π_V: penalty term associated with the violation of the upper voltage limit V_{max} or the lower voltage limit V_{min}, with a penalty coefficient ρ_V assigned by the user in such a way that the penalty is significantly higher than the terms of f_{ref}, without being excessive (otherwise, only feasible solutions would be accepted during the optimization problem, going against the goal of the metaheuristics to open the search space by also accepting penalized cases):

$$\pi_V = \sum_{k=1}^{K}\sum_{h=1}^{H} \rho_V \max\{V_{kh} - V_{max}, V_{min} - V_{kh}, 0\}; \tag{3}$$

- π_R: penalty term associated with the total energy injected at the supply point in the cases of reverse power flow from the low voltage to the medium voltage network that can impact the voltage regulation, as well as the protection systems [30–32]:

$$\pi_R = \sum_{h=1}^{H} \rho_R \max\{-P_{0h}, 0\}\Delta t. \tag{4}$$

2.5. Overall Calculation Procedure

The calculation procedure is composed of two main steps, as shown in Figure 1:

(1) Step A, essentially based on the exploitation of the features of a customized genetic algorithm, and
(2) Step B, where the calculation of potential scheduling for all batteries installed is suggested through a greedy algorithm.

On the basis of the outputs of Step B, the objective function related to that particular set of BESSs is evaluated.

First of all, all the input data are introduced: in particular, the code requires information about the network data, number of nodes K_{BESS} where the BESS can be installed, time horizon (through the number of days N_d) analysed, time discretization (i.e., number of time steps per day N_t), and number of chromosomes N_c for the genetic algorithm. Thanks to the above information, the procedure continues with Step A (planning) and Step B (dispatching). After the calculation of the objective function at the iteration n_G for all the chromosomes, the convergence criterion is checked.

2.5.1. Step A

The initial population $\mathcal{P}_0$ is created and the initial objective function values are collected in the vector $f_{obj}^{(0)}$. The population is composed of chromosomes with the coding shown in Table 2.

Table 2. Codification used in the proposed genetic algorithm.

Gene 1	Gene 2	Gene 3
Number of BESSs installed	Nodes where BESSs are installed	BESS energy capacities (kWh)

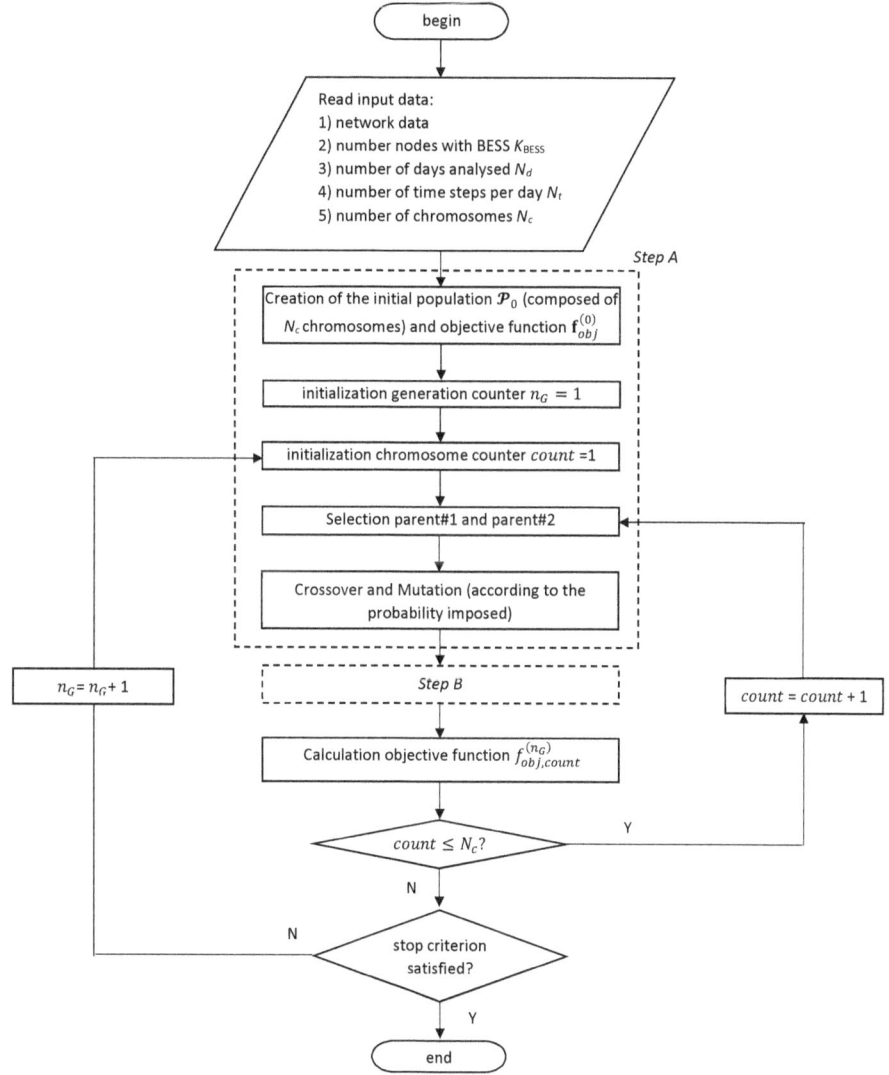

Figure 1. Scheme of the proposed procedure.

It is worth noting that according to the number of BESSs installed (indicated by *gene 1*), the number of elements composing *gene 2* and *gene 3* will vary. This implies that a number of pre-defined rules are needed to handle the genetic operators applied to the chromosomes, to make the creation of successive generations possible.

1. Selection and Crossover

The selection process is based on the application of the biased roulette wheel.

Once the parents' selection is made, the crossover may be applied according to the probability of crossover p_c.

The crossover follows some general rules:

- The number of batteries is inherited from the *parent#1*, and
- The crossover is only applied to the values contained in *gene 2* and *gene 3*.

Due to the particular codification used, some fixed rules have been used to apply crossover in the presence of chromosomes with different characteristics. These rules are clarified with reference to the chromosomes shown in Figure 2.

	gene 1	gene 2	gene 3
chr#1	2	{10,18}	{20,40}
chr#2	4	{5,10,12,14}	{10,20,25,30}

Figure 2. Example of chromosomes.

Different cases do exist:

- If *parent#1* = *chr#1*, the offspring will be that shown in Figure 3a (*offspring#1*), and
- If *parent#1* = *chr#2*, the offspring will be that shown in Figure 3b (*offspring#2*). In this particular case, the number of batteries imposed by *parent#1* is higher than that available in *parent#2*. Therefore, information regarding the remaining nodes and capacities is randomly picked up from a repository containing all the nodes (and related capacities) referring to the current population (e.g., nodes {15, 24} and capacities {5, 10} in *offspring#2*).

offspring#1	2	{5,10}	{10,20}

(**a**) Codification of *offspring#1*

offspring#2	4	{10,18,15,24}	{20,40,5,10}

(**b**) Codification of *offspring#2*

Figure 3. Example of offpring when the parents have different lengths.

2. Mutation

As the crossover operator, the mutation is only applied to the contents of *gene 2* and *gene 3*. This operator is only applied when the value of a random number extracted from a uniform distribution is lower than the probability of mutation p_m. If the mutation is allowed, the values to be substituted are picked up from the repository containing all the nodes and relative capacities of the current population.

2.5.2. Step B

During *Step B*, the procedure calculates the operation setpoints of all the installed batteries. These setpoints are required to calculate the objective function related to the configuration (number of batteries, their positions, and their capacities) specified in every chromosome comprising the population in the generation n_G. The flow chart related to *Step B* is shown in Figure 4.

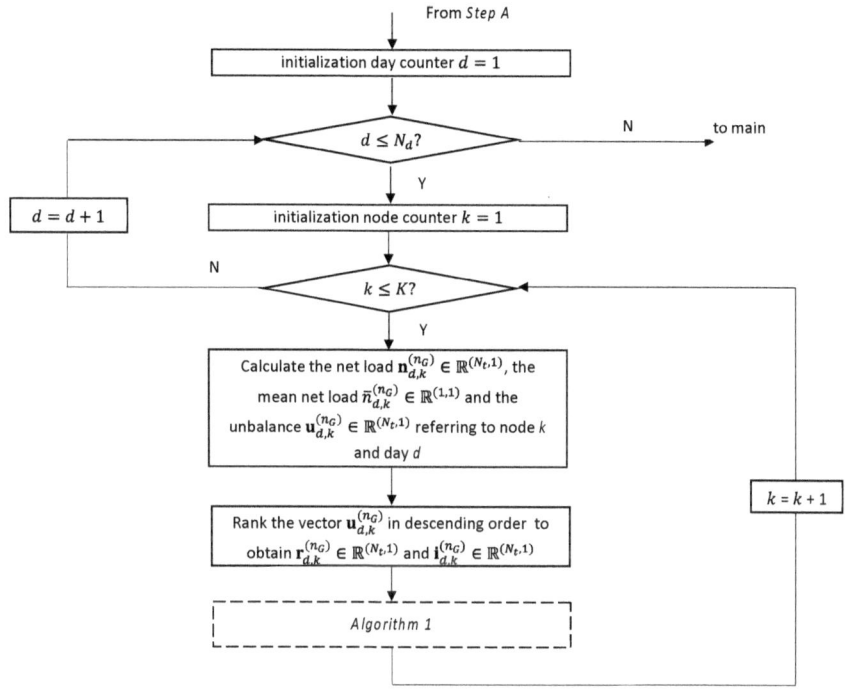

Figure 4. Flowchart related to *Step B*.

For every node $k = 1, \ldots, K$ where the batteries are installed, the net load $\mathbf{n}_{d,k}^{(n_G)} \in \mathbb{R}^{(N_t,1)}$ is calculated as the difference between the load and local generation. With this information, it is possible to calculate the mean value of the nodal net load $\bar{n}_{d,k}^{(n_G)} \in \mathbb{R}^{(1,1)}$ and thus an indicative value of unbalance of the net load with respect to its mean value, i.e., $\mathbf{u}_{d,k}^{(n_G)} \in \mathbb{R}^{(N_t,1)}$, calculated as

$$\mathbf{u}_{d,k}^{(n_G)} = \left| \mathbf{n}_{d,k}^{(n_G)} - \bar{n}_{d,k}^{(n_G)} \right|, \quad k = 1, \ldots, K. \tag{5}$$

The vector $\mathbf{u}_{d,k}^{(n_G)}, k = 1, \ldots, K$, provides information regarding the time steps during which the use of the battery system can be useful to level the net load.

On the basis of the value of $\mathbf{u}_{d,k}^{(n_G)}$, the daily time steps are reordered in a descending way, by obtaining the unbalance ranked vector $\mathbf{r}_{d,k}^{(n_G)} \in \mathbb{R}^{(N_t,1)}$ and the corresponding ranked index vector $\mathbf{v}_{d,k}^{(n_G)} \in \mathbb{R}^{(N_t,1)}$: by knowing the time steps when the unbalance is higher, it is possible to apply the scheduling algorithm by starting from the "most critical" time steps.

The pseudo-code of the scheduling algorithm is shown in Figure 5. *Algorithm 1* tries to set the best option for the current time instant, which is according to the order of solving, more critical with respect to the next time instants, and pushes the battery constraint violations towards the lowest priority time steps. First of all, the set-points for a single battery (collected in the vector called $\mathbf{p}_{BESS}^{(n_G)}$) are instructed, starting from the value of unbalance $\mathbf{u}_{d,k}^{(n_G)}$. The feasibility of that desired pattern depends on the maximum exploitable storage level of the battery and its maximum charging power, called $P_{max}^{(-)}$, as well as its maximum discharging power $P_{max}^{(+)}$. According to the indices value collected in $\mathbf{r}_{d,k}^{(n_G)}$, the algorithm starts from the most critical time step that is the first element of $v_{d,k}^{(n_G)}$ and sets the set-point at that step by respecting the battery's *power constraint*, as shown at line 5.

```
Algorithm 1: Battery Scheduling

1)  p(nG)BESS,k = [NaN, NaN, NaN, ..., NaN]1*T
2)  for i=1 to T do
3)      j = v(nG)d,k,i
4)      if r(nG)d,k,i > 0 then    //Disharge
5)          p(nG)BESS,k,j = min (r(nG)d,k,i, P(-)max )
6)          I = SoC0
7)          for t = 1 to T do
8)              I ← I - ( ω p(nG)BESS,k,t + ( ω -1) [min (SoCmax - I, P(+)max ) ] )
9)              Cf = max (SoCmin - I, 0)
10)             if Cf > 0 then
11)                 p(nG)BESS,k,j ← p(nG)BESS,k,j - p(nG)BESS,k,j
12)                 exit t
13)         end t

14)     if r(nG)d,k,i < 0 then    //Charge
15)         p(nG)BESS,k,j = max (p(nG)BESS,k,j, P(+)max )
16)         I = SoC0
17)         for t = 1 to T do
18)             I ← I - ( ω p(nG)BESS,k,t + (1- ω) [min (I - SoCmin, P(-)max ) ] )
19)             Cf = min (I - SoCmax, 0)
20)             if Cf > 0 then
21)                 p(nG)BESS,k,j ← p(nG)BESS,k,j + p(nG)BESS,k,j
22)                 exit t
23)         end t
24) end i
```

Figure 5. Pseudo-code of the battery scheduling.

Then, an auxiliary variable called I (that stands for *integrator*) is initialized at line 6 with the last value of energy stored in the battery. Following this, the integration operation is executed to check whether the *energy constraint* is respected or not. The parameter ω is defined as follows:

$$\omega = \begin{cases} 0, & p_{BESS,t}^{(n_G)} = NaN \\ 1, & \text{else} \end{cases}.$$

Line 8 indicates that a temporary value for an integral operation equal to the maximum availability state is considered at the time step t. The maximum availability state changes if either a charging or discharging mode is considered: in the discharging mode, the state is the *fully charged state*, whereas, in charging mode, the maximum availability state is the empty one.

Once the integration operator goes beyond the maximum and minimum state of charge (SoC_{max} and SoC_{min}), a correction factor called C_f resets $p_{BESS,j}^{(n_G)}$ and breaks integration execution. This operation is similarly carried out for the charging mode with corresponding signs.

2.6. Selection of the Planning Alternatives

For each scenario, the ranking of the solutions is made on the basis of the objective function (from the best solutions to the worst ones), and the Z top-ranked solutions are selected. The rationale for this selection is that there is no guarantee that the global optimum will be reached from the execution of the metaheuristic, so, taking more than one solution from the ranking, enhances the possibility of having good candidates to compare with the MCDM approach.

The number of planning alternatives is then defined as $A = M \times Z$, that is, with the Z top-ranked solutions for each one of the M scenarios analysed.

Since each planning alternative $a = 1, \ldots, A$ exhibits a different performance according to the scenario, $f_P(a, m)$ is the value of the objective function f_P defined in (2), evaluated for the alternative a when the scenario m occurs. The objective function values are then arranged into a matrix with A rows (planning alternatives) and M columns (scenarios).

The MCDM approach is based on the application of decision theory criteria to the A planning alternatives by considering the M scenarios. In the framework of the decision theory concepts, several criteria can be applied to select the optimal planning alternative, taking into account that each scenario m has a probability of occurrence p_m.

2.6.1. Criterion of Minimum Expected Cost

The criterion of the minimum expected cost attempts to minimize the costs [33]. The optimal planning alternative a^*_{ec} is the one that minimizes the expected cost EC:

$$a^*_{ec} = \arg\min_{a=1,\ldots,A} \{EC(a)\}, \tag{6}$$

where the expected cost $EC(a)$ for each alternative $a = 1, \ldots, A$ is determined as

$$EC(a) = \sum_{m=1}^{M} p_m f_p(a, m). \tag{7}$$

The assignment of the probabilities of occurrence is crucial; when the equal likelihood criterion is adopted [34], each scenario has the same probability of occurrence.

2.6.2. Criterion of Minimax Weighted Regret

The criterion of minimax weighted regret attempts to minimize the regret corresponding to the worst case [34]. For a given scenario m, it is possible to identify the best planning alternative as the one corresponding to the lowest cost; if a planning alternative different from the optimal one is chosen, a greater cost will be experienced and, therefore, the regret can be calculated. Let us consider the scenario m, where the best planning alternative a^*_m is

$$a^*_m = \arg\min_{a=1,\ldots,A} \{f_p(a, m)\} \tag{8}$$

and, when scenario m occurs and the alternative a different from a^*_m is chosen, the regret can be quantified as

$$R(a, m) = f_p(a, m) - f_p(a^*_m, m). \tag{9}$$

Considering the probability of occurrence of each scenario, the weighted regret $R_w(a, m)$ is determined as

$$R_w(a, m) = p_m R(a, m). \tag{10}$$

According to the criterion of minimax weighted regret, the optimal planning alternative a^*_{wr} is the one that minimizes the maximum weighted regret, that is,

$$a^*_{wr} = \arg\min_{a=1,\ldots,A} \left\{ \max_{m=1,\ldots,M} (R_w(a, m)) \right\}. \tag{11}$$

As in the case of the criterion of Section 2.6.1, the equal likelihood criterion can be adopted [33].

2.6.3. "Optimist" and "Pessimist" Criterion

When the "optimist" criterion is applied [28,31], for each planning alternative, the best value of the costs (i.e., the minimum value) over the possible scenarios is selected and, then, the selected planning alternative a^*_{opt} is the one that minimizes the cost corresponding to the best possible outcome for each scenario, that is,

$$a^*_{opt} = \arg\min_{a=1,\dots,A} \left\{ \min_{m=1,\dots,M} f_p(a,m) \right\}. \tag{12}$$

Conversely, the "pessimist" criterion [28,31] attempts to minimize the worst outcome of the planning alternatives. Therefore, the worst value of the costs (i.e., the maximum value) of each alternative over the possible scenarios is selected and, then, the selected planning alternative a^*_{pes} is the one that minimizes the worst outcome, that is,

$$a^*_{pes} = \arg\min_{a=1,\dots,A} \left\{ \max_{m=1,\dots,M} f_p(a,m) \right\}. \tag{13}$$

In addition, an "optimist"-"pessimist" criterion can be considered as a mixed approach. In this case, both the worst and best outcome of each alternative are considered and these values are weighted by a proper factor $\alpha \in [0,1]$:

$$a^*_{opt-pes} = \arg\min_{a=1,\dots,A} \left\{ \alpha \min_{m=1,\dots,M} f_p(a,m) + (1-\alpha) \max_{m=1,\dots,M} f_p(a,m) \right\}. \tag{14}$$

When applying the "optimist", the "pessimist", and the "optimist"-"pessimist" criteria, the decision does not depend on the probabilities of the scenarios.

3. Case Study Application and Results

3.1. Network Data

The application presented in this paper is based on a rural LV network with 22 nodes, including the slack node (Figure 6). The network is radial and is supplied by the MV system through an MV/LV transformer (not represented in the figure). The network contains mainly residential and agricultural customers. The number of nodes considered for BESS installation is $K_{BESS} = 5$. The BESS data are shown in Table 3. For discretization of the BESS energy capacity, the number of levels used is $\Lambda = 47$, with BESS energy capacities considered to lie in the range 3 ÷ 49 kWh, and the energy to power ratio is equal to 2, as indicated in Section 2.3 [29]. The total number of alternative combinations of BESS sizes is $S = \Lambda^{K_{BESS}} = 47^5 = 2.29 \times 10^8$. Such a number is practically intractable with an exhaustive search process. In this situation, the use of a metaheuristic algorithm to identify the solution of the planning problem is justified.

Table 3. Battery Energy Storage System (BESS) capacity data.

Minimum Energy Capacity (kWh)	Maximum Energy Capacity (kWh)	Discretization Steps	Energy to Power Ratio
3	49	47	2

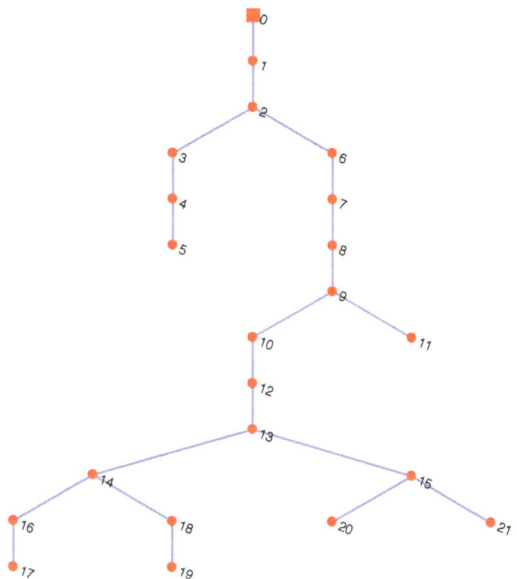

Figure 6. Rural LV network.

3.2. Input Data

Without loss of generality, the reference period T_{ref} is assumed to be one year, and the planning time horizon is assumed to be 15 years (i.e., $N_H = 15$).

The batteries chosen are Lithium Ion (type Nickel Manganese Cobalt): the cost used in this paper is the one expected in 2020 and equal to 167 $/kWh, whereas the inverter cost is equal to 50 $/kW. The operation and maintenance cost is equal to 1% of the investment cost per year. The BESS lifetime has been considered to be equal to 13.4 years [29]. The BESS degradation has been modeled as a reduction of the energy capacity of the battery, equal to 2.4% per year [35].

For execution of the genetic algorithm, the probability of crossover is imposed as $p_c = 0.75$, and the probability of mutation is set to $p_m = 0.05$.

3.2.1. Loads

Regarding the calculation of the load profiles, it takes into account the real contractual delivery powers of each load. Two nodes are considered as commercial consumers, while the other ones refer to residential customers. For the profiles, due to a lack of complete information from the actual profiles, relevant (residential and commercial) profiles have been taken from an open-source dataset (OpenEI) [36], normalized and multiplied by the nominal powers.

3.2.2. PV Generation

The considered PV production, in the simulation, is calculated based on yearly solar irradiance and the nominal installed power [37]. Yearly solar irradiance data was obtained through collected data from a third party weather service provider [38] for the specific location of the case study with the coordinates of 46.4746° N, 11.2479° E.

3.2.3. EV Relevant Information

EVs are considered in some scenarios, through the load profile caused by their charging. Within this analysis, the number of EVs that determines the EV charging profile is set to 10 at the beginning of the scenarios with EV and that number gradually increases by the rate of one EV per year. The nominal power of the EV chargers is considered to be 3.6 kW, i.e., a standard power level of charging stations, and charging events mainly occur during the night.

3.2.4. Energy Price Evolution

Energy price evolutions increase and decrease linearly, with the granularity of one year along entire scenario horizons.

3.3. Definition of the Scenarios

The scenarios are defined by assuming the following entries:

(a) *Electricity prices*: $M_P = 2$ trends of variation are considered, namely, with a 50% increase and 20% reduction of the prices in the planning time horizon of 20 years. Linear variations of the prices are assumed during the years;
(b) *Diffusion of the local generation*: $M_{DG} = 2$ trends of variation of the local generation are considered, with a 20% and 50% increase of the energy production from local generation in the planning time horizon of 20 years. Linear variations of the local generation diffusion are assumed during the years;
(c) *Diffusion of charging points for electric vehicles*: $M_{EV} = 2$ trends are assumed, in which (i) no parking lot and no battery charging stations will be built, and (ii) one parking lot considering 10 active (i.e., with an EV connected for charging purposes) charging points at node 13 is assumed, with the increase rate of the number of active charging points of 1 per year.

3.4. Calculation of the Objective Function

The objective function has been evaluated for each planning alternative and scenario. The best three solutions for each one of the eight scenarios have been calculated for all the scenarios to provide the starting matrix to carry out the calculations based on decision theory. The number of alternatives is then 24. Table 4 shows the objective function values: it is worth noting that, with respect to scenario 2, the values of the objective function for alternatives 3, 8, and 12 are 0.5135, 0.5136, and 0.5140, respectively. Scenarios 5–8 (associated with a large deployment of EVs) exhibit higher values of the objective function. Table 5 reports the technical data of the alternatives, including the number of BESSs installed, the nodes where each BESS is located, and the BESS energy capacity at each node. The solutions include one to three BESSs. Looking at the BESS locations, it can be seen that, in three cases, the slack node is chosen to install a BESS. In this way, the BESS acts to limit the reverse power flows. In the alternative number 23, node 8 is selected two times, because the energy capacity of the BESS in one installation reaches the limits, and the procedure did not find a better alternative to place the BESS in another node.

Table 4. Values of the objective function for each planning alternative and for each scenario ($\times 10^{12}$). Values in bold indicate the best solution for each scenario.

Alternative	Scenarios							
	1	2	3	4	5	6	7	8
1	0.373	0.600	2.44	3.39	5.98	8.31	16.7	23.2
2	0.371	0.515	1.78	2.47	5.84	8.10	16.3	22.6
3	0.371	**0.514**	1.78	2.47	5.83	8.10	16.3	22.6
4	0.381	0.529	2.07	2.88	5.15	7.15	16.5	22.9
5	0.351	0.543	1.78	2.47	5.46	7.58	16.2	22.5
6	0.393	0.545	2.09	2.90	5.55	7.71	16.9	23.5
7	0.412	0.572	1.38	1.92	5.65	7.84	14.2	19.8
8	0.370	0.514	1.39	1.93	5.82	8.08	14.6	20.3
9	0.431	0.598	1.39	1.92	6.00	8.33	**13.5**	**18.7**
10	0.366	0.601	1.49	2.07	5.97	8.29	14.8	20.6
11	0.366	0.628	1.45	2.01	5.97	8.28	14.3	19.9
12	0.350	0.514	1.78	2.10	5.97	8.29	15.0	20.8
13	0.406	0.564	2.20	3.06	5.22	7.24	17.0	23.6
14	0.433	0.601	1.73	2.41	5.18	8.31	15.7	21.8
15	0.432	0.600	1.74	2.42	5.35	8.31	15.8	22.0
16	0.402	0.559	2.19	3.04	5.17	7.17	16.9	23.4
17	0.404	0.561	2.12	2.94	5.19	7.21	16.1	22.3
18	0.402	0.559	2.20	3.06	5.49	7.62	17.0	23.6
19	0.378	0.525	1.75	2.43	**5.13**	**7.12**	16.5	23.0
20	0.404	0.561	**1.28**	**1.78**	5.24	7.28	14.7	20.4
21	0.431	0.598	1.91	2.65	5.96	8.28	16.5	22.9
22	**0.348**	0.567	1.46	2.03	5.84	8.11	14.9	20.6
23	0.351	0.600	1.80	2.50	5.96	8.28	15.8	21.9
24	0.366	0.602	2.20	3.05	5.99	8.31	17.0	23.6

Table 5. Technical data of the planning alternatives.

Alternative	Number of BESSs	Nodes	Energy Capacity (kWh)
1	1	14	21
2	2	0, 18	23, 49
3	2	15, 18	39, 49
4	2	10, 11	47, 12
5	2	18, 20	36, 40
6	2	20, 13	28, 11
7	3	14, 17, 10	21, 42, 9
8	2	11, 18	3, 47
9	3	7, 11, 17	34, 36, 19
10	2	9, 17	28, 40
11	2	17, 16	38, 21
12	2	8, 18	24, 34
13	1	10	29
14	1	19	38
15	1	19	21
16	3	10, 1, 8	36, 6, 18
17	2	10, 7	32, 28
18	2	10, 12	15, 43
19	2	13, 10	42, 16
20	2	10, 17	28, 45
21	3	17, 8, 0	12, 49, 16
22	3	13, 17, 8	13, 41, 20
23	3	8, 17, 8	16, 23, 49
24	3	9, 8, 0	15, 49, 28

3.5. Decision Theory-Based Assessment of the Planning Alternatives and Scenarios

Several cases have been considered, with different values of probability of occurrence assigned to each scenario (Table 6). In Case 1, all scenarios have the same probability of occurrence and, then, the equal likelihood criterion is considered. Cases 2 and 3 consider that it is more probable that the scenarios with higher and lower price increases will occur, respectively. Cases 4 and 5 are focused on the deployment of EVs. In particular, in Case 4, the scenarios with the large deployment of EVs are weighted more, while in Case 5, the scenarios with no deployment of EVs are weighted more. In Case 6, the probabilities are higher for scenarios with a higher price increase and higher PV installation. Finally, Case 7 weights the scenarios with a lower price increase and large deployment of EVs more.

Table 6. Values of scenario probabilities.

	Case 1	Case 2	Case 3	Case 4	Case 5	Case 6	Case 7
Scenario 1	0.125	0.05	0.20	0.05	0.20	0.05	0.15
Scenario 2	0.125	0.20	0.05	0.05	0.20	0.15	0.05
Scenario 3	0.125	0.05	0.20	0.05	0.20	0.05	0.15
Scenario 4	0.125	0.20	0.05	0.05	0.20	0.25	0.05
Scenario 5	0.125	0.05	0.20	0.20	0.05	0.05	0.25
Scenario 6	0.125	0.20	0.05	0.20	0.05	0.15	0.05
Scenario 7	0.125	0.05	0.20	0.20	0.05	0.05	0.25
Scenario 8	0.125	0.20	0.05	0.20	0.05	0.25	0.05

3.5.1. Application of the Criterion of the Minimum Expected Cost

Table 7 reports the expected costs for each alternative in all cases considered. From the values of the expected costs, the selected planning alternative is Alternative 9 for cases 1–4 and 6–7. Alternative 9 is characterized by lower costs in scenarios 7 and 8 (generally associated with high values of costs); as such, it is often the selected alternative. The criterion of the minimum expected cost provides Alternative 20 as the preferred one in Case 5, when scenarios 1–4 are more weighted.

Table 7. Expected costs ($\times 10^{12}$). Values in bold indicate the best solution for each case.

Alternative	Case 1	Case 2	Case 3	Case 4	Case 5	Case 6	Case 7
1	7.61	8.36	6.87	11.2	4.07	9.24	7.86
2	7.24	7.94	6.53	10.8	3.67	8.76	7.53
3	7.24	7.94	6.53	10.8	3.66	8.76	7.53
4	7.20	7.91	6.50	10.6	3.76	8.81	7.46
5	7.12	7.82	6.42	10.6	3.62	8.66	7.40
6	7.45	8.18	6.73	11.0	3.87	9.09	7.73
7	6.47	7.10	5.84	9.71	3.23	7.77	6.75
8	6.63	7.28	5.99	10.0	3.28	7.96	6.92
9	**6.35**	**6.97**	**5.73**	**9.51**	3.19	**7.55**	**6.61**
10	6.78	7.45	6.11	10.2	3.39	8.13	7.06
11	6.62	7.27	5.97	9.92	3.32	7.92	6.89
12	6.86	7.50	6.21	10.3	3.45	8.21	7.15
13	7.41	8.14	6.69	10.9	3.90	9.08	7.67
14	7.02	7.78	6.27	10.5	3.58	8.54	7.20
15	7.08	7.83	6.34	10.6	3.61	8.60	7.29
16	7.36	8.07	6.64	10.8	3.87	9.01	7.61
17	7.11	7.80	6.41	10.5	3.75	8.67	7.35
18	7.48	8.21	6.75	11.0	3.93	9.14	7.75
19	7.11	7.80	6.41	10.6	3.61	8.69	7.39
20	6.45	7.08	5.82	9.71	**3.18**	7.79	6.73
21	7.40	8.12	6.68	11.0	3.80	8.96	7.68
22	6.73	7.39	6.07	10.1	3.35	8.09	7.01
23	7.15	7.86	6.45	10.7	3.65	8.63	7.43
24	7.64	8.39	6.89	11.3	3.99	9.28	7.91

3.5.2. Application of the Minimax Weighted Regret Criterion

Table 8 reports the maximum weighted regrets for each alternative in all cases considered.

Table 8. Maximum weighted regrets (×10^{12}). Values in bold indicate the best solution for each case.

Alternative	Case 1	Case 2	Case 3	Case 4	Case 5	Case 6	Case 7
1	0.560	0.896	0.644	0.896	0.322	1.120	0.805
2	0.486	0.778	0.560	0.778	0.195	0.973	0.700
3	0.486	0.778	0.560	0.778	0.194	0.972	0.700
4	0.534	0.854	0.615	0.854	0.219	1.068	0.768
5	0.483	0.773	0.556	0.773	0.193	0.966	0.696
6	0.605	0.967	0.696	0.967	0.242	1.209	0.870
7	**0.136**	**0.218**	**0.157**	**0.218**	**0.055**	0.273	**0.196**
8	0.206	0.330	0.238	0.330	0.083	0.413	0.297
9	0.151	0.241	0.174	0.241	0.060	**0.181**	0.217
10	0.242	0.386	0.278	0.386	0.097	0.483	0.347
11	0.154	0.247	0.177	0.247	0.062	0.308	0.221
12	0.270	0.431	0.310	0.431	0.108	0.539	0.388
13	0.617	0.988	0.710	0.988	0.255	1.235	0.888
14	0.391	0.626	0.450	0.626	0.156	0.782	0.562
15	0.413	0.661	0.475	0.661	0.165	0.826	0.593
16	0.596	0.953	0.686	0.953	0.253	1.191	0.857
17	0.457	0.732	0.527	0.732	0.233	0.915	0.658
18	0.612	0.979	0.705	0.979	0.255	1.224	0.881
19	0.537	0.859	0.618	0.859	0.215	1.074	0.772
20	0.212	0.339	0.244	0.339	0.085	0.424	0.305
21	0.527	0.843	0.607	0.843	0.211	1.054	0.759
22	0.243	0.388	0.279	0.388	0.097	0.486	0.349
23	0.406	0.650	0.468	0.650	0.162	0.812	0.585
24	0.616	0.985	0.710	0.985	0.254	1.232	0.887

From the values of the maximum weighted regrets, the selected planning alternative is Alternative 7, for cases 1–5 and 7. The values of the objective function in scenarios with a large deployment of EVs influence the results also for this criterion, as in 3.5.1.

3.5.3. Application of the "Optimist-Pessimist" Criterion

The "optimist-pessimist" criterion was applied, considering different values of the weighting factor α ranging from 0 to 1. The case with $\alpha = 0$ corresponds to the application of the "pessimist" criterion, while the case with $\alpha = 1$ is equal to the application of the "optimistic" criterion. The results are reported in Table 9.

Table 9. Selected planning alternatives obtained by applying the "optimist-pessimist" criterion.

α	Selected Planning Alternative
0.0	9
0.1	9
0.2	9
0.3	9
0.4	9
0.5	9
0.6	9
0.7	9
0.8	9
0.9	9
1.0	22

From these results, Alternative 9 is the selected alternative for all the values of the weighting factor, with the exception of the case with $\alpha = 1$. The "optimistic" criterion ($\alpha = 1$) selects Alternative

22, which is associated with the lowest value of the objective function; for the other values of the weighting factor α, higher values of the objective function in scenarios 7–8 influence the results.

4. Conclusions

This paper has presented a novel procedure that combines planning and scheduling of the BESSs installed in an LV grid. In this way, BESS siting and sizing is carried out with the support of a specific assessment of the system operation. The proposed approach combines the properties of the metaheuristics used to search for solutions in a wide space (for the creation of planning alternatives) and the fast calculation of the greedy procedure that allows a viable solution to be found for BESS scheduling. This approach makes it possible to overcome the limitation due to the intractable total number of combinations of BESS sizes, simultaneously handling the non-trivial operational aspects linked to variations in time of the power flows in the network and the BESS scheduling. Furthermore, the uncertainty regarding future scenarios has been handled with a decision theory-based method.

Alternatives 7 and 9 have emerged as the most promising ones when using the decision theory criteria. In particular, Alternative 9 has been preferred by the expected costs and the "optimist-pessimist" criteria, in quite a robust way, over most of the cases with weighted scenarios. Alternative 7 has been preferred by the minimax weighted regrets criterion, again over most of the cases with weighted scenarios. In these two alternatives, three nodes are chosen for BESS installation (with node 17 in common) and the BESS sizes in these nodes are intermediate with respect to the minimum and maximum energy capacity limits used. For a problem of this kind, there can be no guarantee that a globally optimal solution has been reached. For this reason, the creation of meaningful scenarios taking into account the local conditions of loads and generation becomes a fundamental aspect that can be successfully addressed in the case of relatively small electricity communities, where the energy community manager may have an easier view on the decision variables involved in the system operation and planning with respect to what occurs in larger systems. To further develop scenarios of local loads and generation, future research will consider the deployment of other electrification technologies, like heat pumps, that can significantly impact the demand; it will also consider the inclusion of three-phase or single-phase PV systems with storage to take into account the trend of the prosumers to install storage systems in their local plants with the aim of increasing self-sufficiency.

Author Contributions: All the authors contributed equally to this work. All authors have read and agreed to the published version of the manuscript.

Funding: This research received no external funding.

Conflicts of Interest: The authors declare no conflict of interest.

References

1. Zidar, M.; Georgilakis, P.S.; Hatziargyriou, N.D.; Capuder, T.; Škrlec, D. Review of energy storage allocation in power distribution networks: Applications, methods and future research. *IET Gener. Transm. Distrib.* **2016**, *10*, 645–652. [CrossRef]
2. Boicea, V.A. Energy Storage Technologies: The Past and the Present. *Proc. IEEE* **2014**, *102*, 1778–1794. [CrossRef]
3. Chen, S.; Zhang, T.; Gooi, H.B.; Masiello, R.D.; Katzenstein, W. Penetration rate and effectiveness studies of aggregated BESS for frequency regulation. *IEEE Trans. Smart Grid* **2016**, *7*, 167–177. [CrossRef]
4. Brogan, P.V.; Best, R.J.; Morrow, D.J.; McKinley, K.; Kubik, M.L. Effect of BESS response on frequency and RoCof during underfrequency transients. *IEEE Trans. Power Syst.* **2019**, *34*, 575–583. [CrossRef]
5. Brivio, C.; Mandelli, S.; Merlo, M. Battery energy storage system for primary control reserve and energy arbitrage. *Sustain. Energy Grids Netw.* **2016**, *6*, 152–165. [CrossRef]
6. Mosca, C.; Arrigo, F.; Mazza, A.; Bompard, E.; Carpaneto, E.; Chicco, G.; Cuccia, P. Mitigation of Frequency Stability Issues in Low Inertia Power Systems using Synchronous Compensators and Battery Energy Storage Systems. *IET Gener. Transm. Distrib.* **2019**, *13*, 3951–3959. [CrossRef]

7. Zheng, Y.; Dong, Z.Y.; Luo, F.J.; Meng, K.; Qiu, J.; Wong, K.P. Optimal Allocation of Energy Storage System for Risk Mitigation of DISCOs with High Renewable Penetrations. *IEEE Trans. Power Syst.* **2014**, *29*, 212–220. [CrossRef]
8. Qiu, J.; Xu, Z.; Zheng, Y.; Wang, D.; Dong, Z.Y. Distributed generation and energy storage system planning for a distribution system operator. *IET Renew. Power Gener.* **2018**, *12*, 1345–1353. [CrossRef]
9. Saboori, H.; Hemmati, R.; Sadegh Ghiasi, S.M.; Dehghan, S. Energy storage planning in electric power distribution networks—A state-of-the-art review. *Renew. Sustain. Energy Rev.* **2017**, *79*, 1108–1121. [CrossRef]
10. Alegria, E.; Brown, T.; Minear, E.; Lasseter, R.H. CERTS Microgrid Demonstration with Large-Scale Energy Storage and Renewable Generation. *IEEE Trans. Smart Grid* **2014**, *5*, 937–943. [CrossRef]
11. Zheng, Y.; Zhao, J.; Song, Y.; Luo, F.; Meng, K.; Qiu, J.; Hill, D.J. Optimal Operation of Battery Energy Storage System Considering Distribution System Uncertainty. *IEEE Trans. Sustain. Energy* **2018**, *9*, 1051–1060. [CrossRef]
12. Hussain, A.; Bui, V.-H.; Kim, H.M. Impact Analysis of Demand Response Intensity and Energy Storage Size on Operation of Networked Microgrids. *Energies* **2017**, *10*, 882. [CrossRef]
13. Kumar, A.; Meena, N.K.; Singh, A.R.; Deng, Y.; Kumar, P. Strategic integration of battery energy storage systems with the provision of distributed ancillary services in active distribution systems. *Appl. Energy* **2019**, *253*, 113503. [CrossRef]
14. Sedghi, M.; Ahmadian, A.; Aliakbar-Golkar, M. Optimal Storage Planning in Active Distribution Network Considering Uncertainty of Wind Power Distributed Generation. *IEEE Trans. Power Syst.* **2016**, *31*, 304–316. [CrossRef]
15. Ahmadian, A.; Sedghi, M.; Aliakbar-Golkar, M. Fuzzy Load Modeling of Plug-in Electric Vehicles for Optimal Storage and DG Planning in Active Distribution Network. *IEEE Trans. Veh. Technol.* **2017**, *66*, 3622–3631. [CrossRef]
16. Wong, L.A.; Ramachandaramurthy, V.K.; Taylor, P.; Ekanayake, J.B.; Padmanaban, S. Review on the optimal placement, sizing and control of an energy storage system in the distribution network. *J. Energy Storage* **2019**, *21*, 489–504. [CrossRef]
17. Andreotti, A.; Carpinelli, G.; Mottola, F.; Proto, D.; Russo, A. Decision Theory Criteria for the Planning of Distributed Energy Storage Systems in the Presence of Uncertainties. *IEEE Access* **2018**, *6*, 62136–62151. [CrossRef]
18. Tang, Z.; Liu, J.; Liu, Y.; Huang, Y.; Jawad, S. Risk awareness enabled sizing approach for hybrid energy storage system in distribution network. *IET Gener. Transm. Distrib.* **2019**, *13*, 3814–3822. [CrossRef]
19. Pavić, I.; Luburić, A.; Pandžić, H.; Capuder, T.; Andročec, I. Defining and Evaluating Use Cases for Battery Energy Storage Investments: Case Study in Croatia. *Energies* **2019**, *12*, 376. [CrossRef]
20. Yunusov, T.; Frame, D.; Holderbaum, W.; Potter, B. The impact of location and type on the performance of low-voltage network connected battery energy storage systems. *Appl. Energy* **2016**, *165*, 202–213. [CrossRef]
21. Giannitrapani, A.; Paoletti, S.; Vicino, A.; Zarrilli, D. Optimal Allocation of Energy Storage Systems for Voltage Control in LV Distribution Networks. *IEEE Trans. Smart Grid* **2017**, *8*, 2859–2870. [CrossRef]
22. Crossland, A.F.; Jones, D.; Wade, N.S. Planning the location and rating of distributed energy storage in LV networks using a genetic algorithm with simulated annealing. *Int. J. Electr. Power Energy Syst.* **2014**, *59*, 103–110. [CrossRef]
23. Jannesar, M.R.; Sedighi, A.; Savaghebi, M.; Guerrero, J.M. Optimal placement, sizing, and daily charge/discharge of battery energy storage in low voltage distribution network with high photovoltaic penetration. *Appl. Energy* **2018**, *226*, 957–966. [CrossRef]
24. Fortenbacher, P.; Ulbig, A.; Andersson, G. Optimal Placement and Sizing of Distributed Battery Storage in Low Voltage Grids Using Receding Horizon Control Strategies. *IEEE Trans. Power Syst.* **2018**, *33*, 2383–2394. [CrossRef]
25. Khalid Mehmood, K.; Khan, S.U.; Lee, S.; Haider, Z.M.; Rafique, M.K.; Kim, C. Optimal sizing and allocation of battery energy storage systems with wind and solar power DGs in a distribution network for voltage regulation considering the lifespan of batteries. *IET Renew. Power Gener.* **2017**, *11*, 1305–1315. [CrossRef]
26. Babacan, O.; Torre, W.; Kleissl, J. Optimal allocation of battery energy storage systems in distribution networks considering high PV penetration. In Proceedings of the 2016 IEEE Power and Energy Society General Meeting (PESGM), Boston, MA, USA, 17–21 July 2016.

27. Carpinelli, G.; Mottola, F.; Proto, D.; Russo, A.; Varilone, P. A Hybrid Method for Optimal Siting and Sizing of Battery Energy Storage Systems in Unbalanced Low Voltage Microgrids. *Appl. Sci.* **2018**, *8*, 455. [CrossRef]
28. French, S. *Decision Theory, an Introduction to the Mathematics of Rationality*; Ellis Horwood: Chichester, UK, 1989.
29. IRENA. *Electricity Storage and Renewables: Costs and Markets to 2030*; October 2017; ISBN 978-92-9260-038-9. Available online: https://www.irena.org/publications/2017/Oct/Electricity-storage-and-renewables-costs-and-markets (accessed on 17 December 2019).
30. Von Appen, J.; Braun, M.; Stetz, T.; Diwold, K.; Geibel, D. Time in the sun: The challenge of high PV penetration in the German electric grid. *IEEE Power Energy Mag.* **2013**, *11*, 55–64. [CrossRef]
31. Mortazavi, H.; Mehrjerdi, H.; Saad, M.; Lefebvre, S.; Asber, D.; Lenoir, L. A monitoring technique for reversed power flow detection with high PV penetration level. *IEEE Trans. Smart Grid* **2015**, *6*, 2221–2232. [CrossRef]
32. De Carne, G.; Buticchi, G.; Zou, Z.; Liserre, M. Reverse Power Flow Control in a ST-Fed Distribution Grid. *IEEE Trans. Smart Grid* **2018**, *9*, 3811–3819. [CrossRef]
33. Anders, G.J. *Probability Concepts in Electric Power Systems*; John Wiley & Sons: New York, NY, USA, 1990.
34. Miranda, V.; Proenca, L.M. Probabilistic choice vs. risk analysis-conflicts and synthesis in power system planning. *IEEE Trans. Power Syst.* **1998**, *13*, 1038–1043. [CrossRef]
35. Kubiak, P.; Cen, Z.; López, C.M.; Belharouak, I. Calendar aging of a 250 kW/500 kWh Li-ion battery deployed for the grid storage application. *J. Power Sources* **2017**, *372*, 16–23. [CrossRef]
36. OpenEI. Available online: https://openei.org/datasets/files/961/pub/ (accessed on 17 December 2019).
37. Spertino, F.; Corona, F.; Di Leo, P. Limits of Advisability for Master–Slave Configuration of DC–AC Converters in Photovoltaic Systems. *IEEE J. Photovolt.* **2012**, *2*, 547–554. [CrossRef]
38. World Weather Online. Available online: https://www.worldweatheronline.com/ (accessed on 17 December 2019).

© 2019 by the authors. Licensee MDPI, Basel, Switzerland. This article is an open access article distributed under the terms and conditions of the Creative Commons Attribution (CC BY) license (http://creativecommons.org/licenses/by/4.0/).

Article

Planning of Distributed Energy Storage Systems in µGrids Accounting for Voltage Dips

Fabio Mottola [1], Daniela Proto [1], Pietro Varilone [2] and Paola Verde [2,*]

[1] Department of Electrical Engineering and Information Technology, University of Naples Federico II, I-80125 Naples, Italy; fabio.mottola@unina.it (F.M.); daniela.proto@unina.it (D.P.)
[2] Department of Electrical Engineering and Information "Maurizio Scarano", University of Cassino and Southern Lazio, I-03043 Cassino, Italy; varilone@unicas.it
* Correspondence: verde@unicas.it

Received: 3 December 2019; Accepted: 10 January 2020; Published: 14 January 2020

Abstract: This paper deals with the optimal planning of the electrical energy storage systems in the microgrids aimed at cost minimization. The optimization accounts for the compensation of the voltage dips performed by the energy storage systems. A multi-step procedure, at the first step, identifies a set of candidate buses where the installation of a storage device produces the maximum benefit in terms of dip compensation; then, the life cycle costs in correspondence of different alternatives in terms of size and location of the storage systems are evaluated by considering an optimized use of the energy storage systems. The simulations on a medium voltage microgrid allowed validating the effectiveness of the proposed procedure.

Keywords: electrical energy storage systems; voltage dips; power quality; microgrid planning

---

1. Introduction

The role of the systems for storing the electrical energy is gaining more and more importance in the frame of modern power systems. This is thanks to the electrical energy storage systems' (EESSs) ability to provide a number of benefits across multiple levels [1–3]. Focusing on the distribution level of electrical energy systems, the EESSs' benefits are mainly related to the compensation action of the intermittent effects of renewable power sources and to the support to the operation of the network by providing services aimed at regulating voltage levels, at reducing losses, and deferring the investment on the distribution system [4–6]. End-users can also benefit from EESSs through reduction of the cost for the energy purchased as well as for the improvement of power quality (PQ) and reliability [7].

The convenience of the installation of the EESSs is still a critical issue and is strictly related to the number of benefits that can be achieved contemporaneously from their installation [8]. In the case of the microgrids (µGs), cost reduction and PQ improvement are matters of particular interest for both grid operators and end-users.

Cost reduction refers to the ability of the µG owner to increase the share of power produced from renewables inside the µG thus reducing the cost sustained for the imported power, as well as to increase the system efficiency and reduce the network losses. PQ improvement refers to possibility of reducing the level of the PQ disturbances at least up to the ranges admissible for the loads fed by the µG. The typical characteristics of a µG make the objective of the PQ improvement more difficult to attain than in the traditional distribution systems. The network structure, the lengths of the lines, the installed powers and the levels of short circuit power jointly contribute to limiting the PQ robustness of a µG, defined as its intrinsic capacity to maintain assigned disturbance levels when the external conditions change [9,10]. The improvement of the performance of a µG in terms of voltage dips, one of the most critical PQ disturbances, represents an attractive challenge for different reasons. First of all, the voltage dips are the disturbances with serious consequences on processes and activities that could evolve in higher sustained costs. Secondly, these costs are important not only for the industrial users, for which are certainly the most critical [11], but also for commercial and service activities. Loss of production, damages to equipment, halt of processes and data losses are some of the effects of voltage dips, which unavoidably evolve into significant financial losses [12]. Further, the detrimental effects of voltage dips can regard also residential customers, which can suffer for uncomfortable supply service. Also, in the frame of smart grids, most of the new smart technologies are based on electronic devices and control systems, which are particularly sensitive. Finally, it is worth emphasizing that the importance of the improvement of the voltage dip performance is also proved by the fact that most of the regulations on the quality of voltage by the national authorities in Europe started with actions on voltage dips, even if only with wide measurement campaigns like in Italy [13].

In this context, where assuring high voltage dips performance is a primary issue, a possible solution could be given by installing a number of EESSs in the μG, able to fast respond to the occurrence of voltage dips. Clearly, the installation of EESSs is not without drawbacks, particularly those related to the cost of this technology. Thus, EESSs need to be optimally sized by maximizing their benefits while minimizing costs and need to be installed in those nodes considered strategical for the maximization of their value in terms of dip compensation.

In the relevant literature, great effort has been paid to the optimal planning of EESSs in μGs with the objective of the cost reduction. Optimal planning of EESSs in the distribution grids is proposed in [6], based on the minimization of the investment and operation costs. In [14], a planning procedure is proposed, which takes into account the minimization of the cost of energy imported from the external grid while considering voltage support and minimization of network losses. The problem of the optimal siting and sizing of EESSs in μGs is faced in [15] through the minimization of both installation and operation costs, as well as power losses. Minimization of investment and operation costs in distribution networks is considered in the planning tool proposed in [16], which accounts for constraints on bus voltages. Costs of both power losses and EESS' operation are included in the analytical planning tool presented in [17]. In [18], the optimal allocation of EESSs in distribution networks is performed on the basis of a tradeoff among investment and operation costs, technical constraints and network losses. A planning tool is proposed in [19] aimed at the minimization of installation and operation costs while accounting for the network technical constraints in the presence of uncertainties. The mitigation of voltage dips thanks to the installation of the storage devices is analyzed in [20], where a cost-benefit analysis is proposed. To the best of our knowledge, approaches that include the benefits related to voltage dip compensation within the EESS planning has never been proposed in the current technical literature.

In this paper, a minimum cost strategy for the planning of EESSs in a μG is proposed where the cost is avoided thanks to the voltage dip compensation being taken into account. At this aim, the optimal siting and sizing of EESSs in the μG is formulated in terms of a multi-period, non-linear, constrained optimization problem. A multi-step approach is proposed based on the identification of a limited set of candidate buses among the most exposed to the voltage dips. An optimization tool is used to identify the total cost related to the possible design alternatives (size and location of EESSs). The selection of the candidate buses allows optimizing EESSs' installation while reducing computational complexity. The reduced computational burden allows analyzing all the possible design alternatives and choosing the optimal solution as the one corresponding to the maximum benefit. Compared to the existing technical literature, the main contribution of this paper in terms of novelty refers to the inclusion of the voltage dips' cost in the planning of EESSs in μGs.

The paper is organized as follows. The problem formulation of the proposed planning method is reported in Section 2. In Section 3, the method used to identify the set of candidate buses is described. Section 4 reports the optimization tool used to derive the total cost of the planning solutions. Some results related to the application of the proposed method to a case study based on an MV test system are summarized and discussed in Section 5. Our conclusions are drawn in Section 5 together with some comments on future research work.

2. Problem Formulation

The planning problem refers to a three-phase, MV, balanced, µG including residential, industrial and commercial loads. The µG is connected to an upstream grid through the Point of Common Coupling (PCC). The distributed resources including Distributed Generation (DG) units and EESSs are connected to the µG through power converters, and are supposed to be owned by the µG operator.

The contribution of both the DG units and EESSs is the provision of active power to supply part of the loads and, reactive power to support the µG operation. The active power profile of the EESSs (that is, their charging/discharging profile), and the reactive power of both DG units and EESSs are determined on the basis of a minimum cost strategy. More specifically, the strategy is aimed at minimizing the cost sustained by the µG operator for the energy imported from the upstream grid. The size and location of the DG units are assigned whereas, the size and location of the EESSs are identified according to the proposed planning problem based on the minimization, a total cost function. The total cost function includes the installation cost of the EESSs, the operation cost of the µG, and the benefit derived from the voltage dip compensation over the whole planning period.

The planning problem is developed in four steps, as shown in the flow chart in Figure 1. The first step refers to the selection of the candidate nodes, chosen as those resulting among the most vulnerable in terms of exposure to the voltage dips. For this reason, these nodes correspond to the sites of the installation of the EES units with the preferred performance in terms of dip compensation. The procedure for the selection is presented in Section 4.

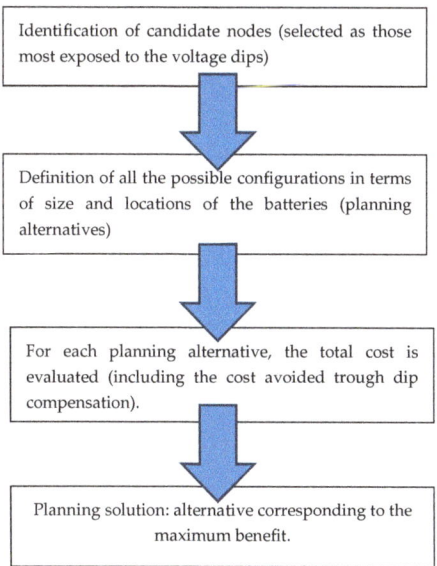

Figure 1. Flow chart of the proposed planning procedure.

The second step refers to the identification of all the possible configurations in terms of size and location for the EESS installation (limiting the installation only to the set of nodes derived from the Step 1). For each of the possible configurations (planning alternatives), the total cost sustained by the µG owner is evaluated according to the procedure detailed in the Section 5 (third step).

Finally, in the fourth step, the best planning alternative is selected as that corresponding to the maximum benefit (BF) derived by the installation of the EESSs. This BF can be evaluated for each alternative by comparing the total cost sustained by the μG's owner in absence of the EESSs ($C_T^{no\ EESS}$) and that resulting when the EESSs are installed (C_T^{EESS}):

$$BF(A_i) = C_T^{no\ EESS}(A_i) - C_T^{EESS}(A_i) \quad i = 1\ldots N_A \tag{1}$$

where N_A is the number of the design alternatives. The total cost evaluated when the EESSs are installed includes the reduction of the cost item related to the voltage dips and of that related to the grid operation thanks to the price arbitrage.

3. Identification of the Set of Buses Most Exposed to the Voltage Dips

A fault in a node can cause a voltage dip in one or more nodes of the electrical network in function of several parameters that determine the electrical power system response to a short circuit. These parameters are linked to the structural characteristics of the network and to the features of the installed components. Examples of such "hardware characteristics" of the electrical system are the network configuration, the installed generator powers, and the size, length and type of the lines.

The numbers of the voltage dips or their frequency in a time horizon, usually one year, is instead linked to the frequency of the causes, the short circuits, which originated them.

The choice of the candidate nodes where the μG operator can place a compensating unit of the voltage dips has been made according to the hardware characteristics of the network that constitutes the μG. In particular, the set of candidate nodes are those nodes that mostly experience the voltage dips whose residual voltage is below a critical value, V_{cr}, regardless the frequency of the faults which originated these voltage dips.

This choice allows ascertaining the sensitiveness of the nodes directly to the effects of the short circuits in any node of the network in terms of voltage dips linked to the characteristics of the network and of the installed components. Consequently, for any possible configuration of the network (radial, meshed, ring) the set of the candidate nodes is different for the same V_{cr}.

Inside a μG, the figure V_{cr} represents the limit value of vulnerability of the loads supplied by the system, that is the minimum value below which the equipment experiences a trip, the most critical effect of a voltage dip.

Theoretically, the value of V_{cr} should vary in function of the specific load fed at every node of the μG. However, a more realistic scenario in a planning stage allows accounting for the vulnerability limits of classes of equipment, as indicated by the International Standards IEC [21]. In the case of equipment supplied by the μG are of Class II or Class III, V_{cr} corresponds to the limit value of 70% or 40% of the declared voltage, respectively, for the voltage dips lasting up to 200 ms.

Summarizing, for every possible configuration of the μG network, the set of candidate busses for the installation of a compensating units is the set of busses which experience the largest number of voltage dips whose residual voltage is below V_{cr}, regardless the frequency of the faults which caused these voltage dips.

The Fault Position Method (FPM) is the most effective tool for the selection of the candidate bus set in the given assumptions.

In fact, for a system with N nodes the FPM allows obtaining the (N×N) matrix of the during fault voltages in any node for short circuit in every node, $[\overline{V}_{df}]$, by means of the following equation for a three-phase fault in every node of the system:

$$[\overline{V}_{df}] = [\overline{E}_{pf}] - [\dot{Z}_{SC}]\left(diag[\dot{Z}_{SC}]\right)^{-1} diag[\overline{E}_{pf}] \tag{2}$$

where $\left[\dot{Z}_{SC}\right]$ is the nodal short circuit impedance matrix, $\left[\overline{E}_{pf}\right]$ is the pre-fault voltage matrix and $\left(diag\left[\dot{Z}_{SC}\right]\right)^{-1}$ is the matrix formed by the inverse of the diagonal elements of the short circuit impedance matrix. If the pre-fault voltages are assumed to be 1 p.u., relation (2) can be written as:

$$\left[\overline{V}_{df}\right] = [ones] - \left[\dot{Z}_{SC}\right]\left(diag\left[\dot{Z}_{SC}\right]\right)^{-1} diag[ones] \quad (3)$$

where [ones] is a matrix full of ones and such that dimension is equal to the dimension of $\left[\dot{Z}_{SC}\right]$.

Any element $\overline{V}_{ik}$ of $\left[\overline{V}_{df}\right]$ with magnitude less than 0.9 p.u. represents a voltage dip in the node i caused by a three-phase short circuit in the node k. For different types of short circuits, similar equations can be drawn [22].

For every network configuration of the μG with N nodes, starting from the matrix $\left[\overline{V}_{df}\right]$ obtained by the FPM, the set of candidate busses is chosen as constituted by the m nodes such that:

$$\left|\overline{V}_{i,k}\right| \leq V_{cr}, \quad i = 1, \ldots, m \text{ and } k = 1, \ldots, N. \quad (4)$$

As mentioned above, the value of V_{cr} in the equation (4) is chosen from the Standard (e.g., [23]).

4. Total Cost Evaluation

For each alternative, the total costs, C_T, related to the inclusion of the EESSs in the μG is given by the sum of the cost of installation, C_{inst}, and replacement, C_{rep}, of the EESSs, the operation costs of the μG, C_{op}, and the cost due to voltage dips.

$$C_T(A_i) = C_{inst}(A_i) + C_{rep}(A_i) + C_{op}(A_i) + C_{vd}(A_i) \quad (5)$$

All the cost items refer to the whole planning period and are detailed in the following sub-sections. In (5), maintenance cost can be also added as a percentage of the installation cost. For ease of notation, in the following equations, the reference to alternative A_i is omitted.

4.1. Installation Cost

The installation cost includes both the cost of the battery and the cost of the power converter interfacing the storage device to the grid. The cost of the battery depends on the energy capacity, whereas the cost of the power converter is related to the rated power. In case of EESSs, the rated energy and power are linked through the nominal C-rate. Then, a unitary installation cost can be provided once the nominal C-rate has been specified. In this case, the installation cost of the EESSs is given by the product of the capacity unitary installation cost, IC_{EESS}, and the battery size. Obviously, the former depends on the considered battery technology and the latter refers to the specific design alternative. The installation cost is then given by:

$$C_{inst} = \sum_{i=1}^{nb} IC_{EESS}\, E^{size}_{EESS_i} \quad (6)$$

where nb is the number of installed EESSs and $E^{size}_{EESS_i}$ is the size of the ith EESS.

4.2. Replacement Cost

Based on the expected battery lifetime, which is given in terms of number of charging/discharging cycles—and that depends on the battery's stress factors, the replacement of the device could be necessary during the considered planning period. The replacement cost is given by:

$$C_{rep} = \sum_{i=1}^{nb} r_i RC_{EESS} E_{EESS_i}^{size} \qquad (7)$$

where RC_{EESS} is the unitary cost for the battery replacement, and r_i indicates the number of times the ith battery needs to be replaced. To evaluate r_i, the battery's lifetime duration must be evaluated and compared to the planning period. The battery's lifetime, in years, (L_b) can be evaluated based on the lifecycle (i.e., life expressed in terms of number of charging/discharging cycles), N_{cycles}, and on the number of cycles per day, $n_{cycles,d}$:

$$L_b = \frac{N_{cycles}}{365 \, n_{cycles,d}} \qquad (8)$$

N_{cycles} represents the total number of cycles the battery can be used before the replacement; it strictly depends on the battery technology and on the way stress factors act [24,25]. Its value is provided by the battery manufacturer, and it is related to specified operation conditions, such as ambient temperature and depth of discharge (DoD). The value of $n_{cycles,d}$ depends on the way the battery is operated, which depends on the operation strategy.

4.3. Operation Cost

The operation cost of the µG refers to the cost of the energy imported from the upstream grid. It is supposed that the µG is not able to sell energy to the upstream grid. The cost of the energy imported implies the evaluation of the overall costs

- sustained by the µG's owner to (i) supply the loads, (ii) charge the EESSs, (iii) compensate for the power losses;
- avoided thanks to the EESS discharged energy.

The planning time horizon includes ny years, each represented by nd_y typical days. Forecasted daily profiles of the power requested by the loads, of the power delivered by the distributed generation, and those of the energy price are known for each typical day of the first year. These profiles are known with reference to all the nt time intervals—of duration Δt—in which each day is divided. The corresponding profiles for the subsequent years are evaluated based on a specified yearly growth. The operation cost for the whole planning period is then provided by the net present value of the sum of the costs sustained at each typical day:

$$C_{op} = \sum_{y=1}^{ny} \frac{1}{(1+a)^{y-1}} \sum_{d=1}^{nd_y} N_{y,d} \left[\sum_{k=1}^{nt} \left(P_{1,k_{(y,d)}} \Delta t \right) Pr_{k_{(y,d)}} \right] \qquad (9)$$

where a is the discount rate, $N_{y,d}$ is the number of days represented by the dth typical day in the year y, $P_{1,k_{(y,d)}}$ and $Pr_{k_{(y,d)}}$ are the power imported from the upstream network and the energy price, respectively, at the kth time interval of the dth typical day of the year y.

In order to evaluate the power imported from the upstream grid, $P_{1,k_{(y,d)}}$, a minimum cost strategy is formulated for the µG, on the basis of a two-stage procedure which allows decoupling the hourly optimization algorithm so reducing the computational burden.

In the first stage, the hourly active power of the EESSs is evaluated. According to the cost minimization approach, the value of the power of each EESS is derived by charging the battery during low-price hours and discharging it during the high price hours, while constraints are imposed on the rated power and energy capacity of the EESSs.

In the second stage an OPF is solved, at each time interval, with respect to the known variables (i.e., specified values of loads' active and reactive powers and active powers of DG units and EESSs) and unknown variables (i.e., active and reactive power at the PCC, reactive powers of both DG units and EESSs.) Coherently with the cost minimization strategy, the OPF minimizes the power imported at the PCC while satisfying the constraints on both the µG and the interfacing converters.

a. 1th Stage: EESS Profile Optimization

The power that can be charged or discharged by the EESSs at each time interval depends on the stored energy available in the battery, on the rated power and on the price of energy corresponding to that interval. A this aim, two sets of time intervals are identified in a day: the first set is related to the highest price—where the EESSs can be discharged to reduce the costs of the imported power—$\Omega_{dch_{(y,d)}}$, the second set is related to the lowest price—where the EESSs can be charged—$\Omega_{ch_{(y,d)}}$. Two different iterative procedures are applied in $\Omega_{dch_{(y,d)}}$ and $\Omega_{ch_{(y,d)}}$ (Figure 2).

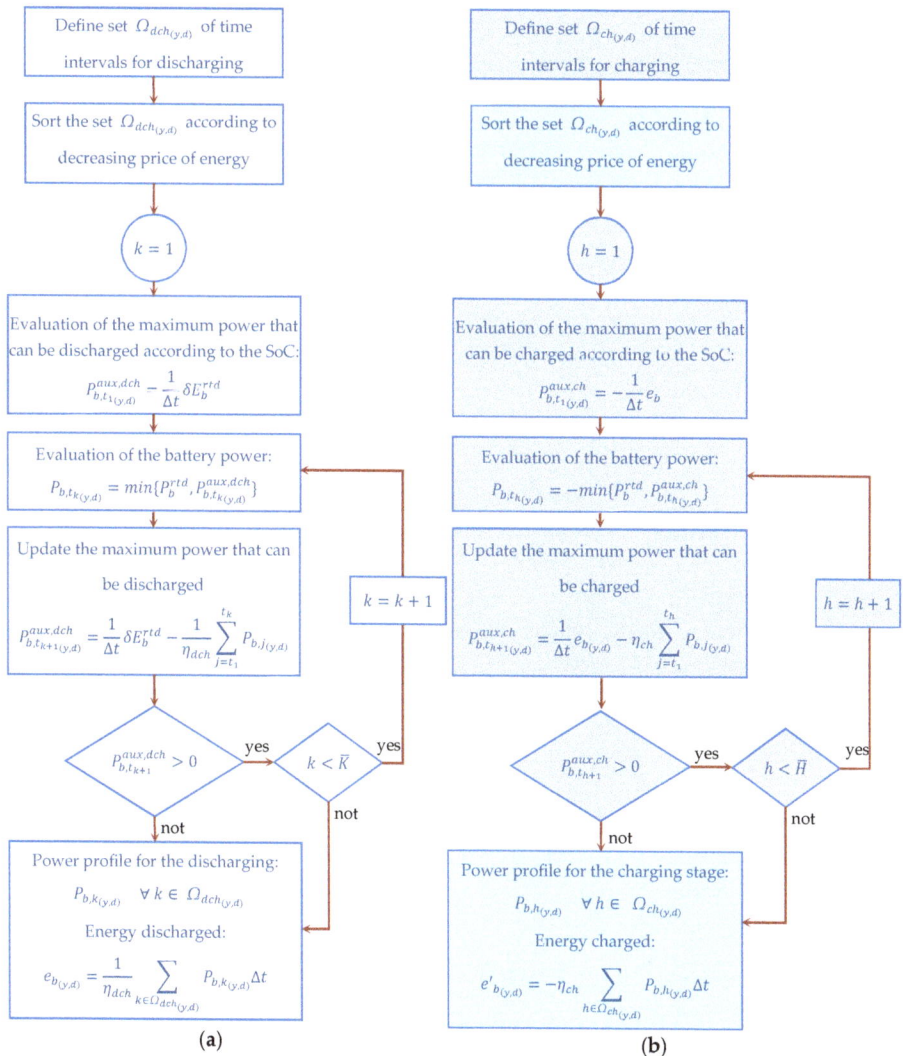

Figure 2. Flow chart of the iterative procedure for the determination of the EESS profile during (a) $\Omega_{dch_{(y,d)}}$ and (b) $\Omega_{ch_{(y,d)}}$.

The iterative procedure that applies to $\Omega_{dch_{(y,d)}}$ allows obtaining the EESS's profile during the discharging stage (Figure 2a). The procedure is based on the identification of the time intervals included in $\Omega_{dch_{(y,d)}}$ and on the identification of the set $\Omega'_{dch_{(y,d)}}$ which includes the time intervals of $\Omega_{dch_{(y,d)}}$ sorted by decreasing price of energy. Then, at each time interval t_k included in $\Omega'_{dch_{(y,d)}}$, with $k = 1,\ldots, \overline{K}$, the power discharged by the battery is identified as the minimum between the EESS rated power (P_b^{rtd}) and the power corresponding to the battery SoC ($P_{b,t_k}^{aux,dch}$):

$$P_{b,t_k_{(y,d)}} = \min\{P_b^{rtd}, P_{b,t_k}^{aux,dch}\} \tag{10}$$

$$t_k \in \Omega'_{dch_{(y,d)}} \qquad y = 1,\ldots, ny, \qquad d = 1,\ldots, nd_y$$

With reference to the first interval in which the battery can be discharged, $k = 1$, it is assumed that the battery is fully charged, thus

$$P_{b,t_1}^{aux,dch} = \frac{1}{\Delta t}\delta E_b^{rtd} \tag{11}$$

where E_b^{rtd} is the rated energy capacity of the battery, and δ is the admissible DoD. Once the battery power is known through (10), the power corresponding to the SoC is updated:

$$P_{b,t_{k+1}}^{aux,dch} = \frac{1}{\Delta t}\delta E_b^{rtd} - \frac{1}{\eta_{dch}}\sum_{j=t_1}^{t_k} P_{b,j_{(y,d)}} \tag{12}$$

where η_{dch} is the discharging efficiency of the EESS. The procedure ends when all of the time intervals in $\Omega'_{dch_{(y,d)}}$ are explored (i.e., $k = \overline{K}$) or $P_{b,t_{k+1}}^{aux,dch} \leq 0$. Outputs of the procedure are the EESSs' power profiles during the discharging stage (i.e., $P_{b,k_{(y,d)}} \forall k \in \Omega_{dch_{(y,d)}}$) and the total energy which can be discharged by the battery:

$$e_{b_{(y,d)}} = \frac{1}{\eta_{dch}}\sum_{k \in \Omega_{dch_{(y,d)}}} P_{b,k_{(y,d)}}\Delta t \tag{13}$$

The iterative procedure, which applies in $\Omega_{ch_{(y,d)}}$ allows obtaining the EESS's profile during the charging stage (Figure 2b). The procedure is based on the identification of the time intervals included in $\Omega_{ch_{(y,d)}}$ and on the identification of the set $\Omega'_{ch_{(y,d)}}$ which includes the time intervals of $\Omega_{ch_{(y,d)}}$ sorted by increasing price of energy. Then, at each time interval t_h included in $\Omega'_{ch_{(y,d)}}$, with $h = 1,\ldots, \overline{H}$, the discharged power of the battery is identified as the minimum between the EESS rated power (P_b^{rtd}) and the power corresponding to the battery SoC ($P_{b,t_h}^{aux,dch}$):

$$P_{b,t_h_{(y,d)}} = -\min\{P_b^{rtd}, P_{b,t_h}^{aux,ch}\} \tag{14}$$

$$t_h \in \Omega'_{ch_{(y,d)}} \qquad y = 1,\ldots, ny, \qquad d = 1,\ldots, nd_y$$

With reference to the first interval in which the battery can be charged, $h = 1$, it is assumed that the battery is empty, thus

$$P_{b,t_1}^{aux,ch} = -\frac{1}{\Delta t}e_{b_{(y,d)}} \tag{15}$$

where $e_{b_{(y,d)}}$ is given by (13). Once the battery power is known through (10), the power corresponding to the SoC is updated:

$$P_{b,t_{h+1}}^{aux,dch} = \frac{1}{\Delta t}e_{b_{(y,d)}} - \eta_{ch}\sum_{j=t_{h1}}^{t_{hh}} P_{b,j_{(y,d)}} \tag{16}$$

Where η_{ch} is the charging efficiency of the EESS. The procedure ends when all of the time intervals in $\Omega'_{ch_{(y,d)}}$ are explored (i.e., $h = \overline{H}$) or $P^{aux,ch}_{b,t_{h+1}} \leq 0$. Outputs of the procedure are the EESS's power profile during the charging stage (i.e., $P_{b,h_{(y,d)}}$ $\forall h \in \Omega_{ch}$) and the total energy, which can be charged by the battery:

$$e'_{b_{(y,d)}} = -\eta_{ch} \sum_{h \in \Omega_{ch_{(y,d)}}} P_{b,h_{(y,d)}} \Delta t \tag{17}$$

In order to make the strategy feasible, it is required that during each day, the energy charged and discharged must be the same:

$$e_{b_{(y,d)}} = e'_{b_{(y,d)}} \tag{18}$$

In the case that (18) is not satisfied, the procedure for the discharging stage (Figure 2) must be repeated by replacing $\delta E^{rtd}_b = e'_{b_{(y,d)}}$.

b. 2nd Stage: Optimal Power Flow

The Optimal Power Flow (OPF) in its general form is formulated as:

$$\min f_{obj}(x) \tag{19}$$

$$\psi_v(x) = 0 \quad v = 1, \ldots, n_{eq} \tag{20}$$

$$\phi_\gamma(x) \leq 0 \quad \gamma = 1, \ldots, n_{ineq} \tag{21}$$

where the minimization of the objective function f_{obj} is subject to equality and inequality constraints, ψ and ϕ applied to the vector x of the optimization variables. Inputs of the OPF are the forecasted values of the power load demand, the DG power production, the energy price and the EESS power. Outputs are the reactive power of the DG units and of the EESSs, and the active and reactive power at the PCC. All these quantities are required to be known at all the k time intervals of each typical day d of year y. In order to minimize the operation costs defined in (9), the objective function is:

$$f_{obj}(x) = \left(P_{1,k_{(y,d)}} \Delta t\right) Pr_{k_{(y,d)}} \quad k = 1, \ldots, nt, \quad y = 1, \ldots, ny, \quad d = 1, \ldots, nd_y \tag{22}$$

The constraints to be satisfied in the OPF include the classical power flow equations:

$$P_{i,k_{(y,d)}} = V_{i,k_{(y,d)}} \sum_{j=1}^{n} V_{j,k_{(y,d)}} \left[G_{i,j} \cos(\delta_{i,j,k_{(y,d)}}) + B_{i,j} \sin(\delta_{i,j,k_{(y,d)}})\right] \tag{23}$$

$$Q_{i,k_{(y,d)}} = V_{i,k_{(y,d)}} \sum_{j=1}^{n} V_{j,k_{(y,d)}} \left[G_{i,j} \sin(\delta_{i,j,k_{(y,d)}}) - B_{i,j} \cos(\delta_{i,j,k_{(y,d)}})\right] \tag{24}$$

$$V_{1,k_{(y,d)}} = V_{spec} \tag{25}$$

$$\delta_{1,k_{(y,d)}} = 0 \tag{26}$$

$$k = 1, \ldots, nt, \quad y = 1, \ldots, ny, \quad d = 1, \ldots, nd_y$$

where n is the number of network buses and, with reference to the time interval k of the day d in the year y, $\delta_{i,j,k_{(y,d)}}$ is the difference between the phase angles at nodes i and j, $P_{i,k_{(y,d)}}$ and $Q_{i,k_{(y,d)}}$ are the active and reactive powers at bus i, $V_{i,k_{(y,d)}}$ is the ith bus voltage amplitude, $G_{i,j}$ and $B_{i,j}$ are the (i, j)-terms of the matrices of the conductance and susceptance, respectively. Constraints (25) and (26) refer to slack bus ($i = 1$), that is the PCC.

The active and reactive powers in (19) and (20) include the powers absorbed by the loads, the power charged/discharged by the EESSs and the power produced by the DG units.

The active and reactive power exchanged at the PCC must comply with the transformer rate, S_{tr}^{rtd}, that is:

$$\left[\left(P_{1,k_{(y,d)}}\right)^2 + \left(Q_{1,k_{(y,d)}}\right)^2\right]^{1/2} \leq S_{tr}^{rtd} \qquad (27)$$

$$k = 1,\ldots, nt, \quad y = 1,\ldots, ny, \quad d = 1,\ldots, nd_y$$

The same applies to the power flowing through the converters interfacing the EESSs and the DG units, which cannot exceed the converter rated power, $S_{DESS,i}^{rtd}$, $S_{DG,i}^{rtd}$:

$$\left[\left(P_{b,i,k_{(y,d)}}\right)^2 + \left(Q_{b,i,k_{(y,d)}}\right)^2\right]^{1/2} \leq S_{DESS,i}^{rtd} \qquad (28)$$

$$i \in \Omega_{DESS}$$

$$\left[\left(P_{dg,i,k_{(y,d)}}\right)^2 + \left(Q_{dg,i,k_{(y,d)}}\right)^2\right]^{1/2} \leq S_{DG,i}^{rtd} \qquad (29)$$

$$i \in \Omega_{DG}$$

$$k = 1,\ldots, nt, \quad y = 1,\ldots, ny, \quad d = 1,\ldots, nd_y$$

where $P_{b,i,k_{(y,d)}}$ has been previously derived in *Step 1*, $Q_{b,i,k_{(y,d)}}$ is the EESS reactive power, and $P_{dg,i,k_{(y,d)}}$ and $Q_{dg,i,k_{(y,d)}}$ are the DG active and reactive powers, Ω_{DESS} and Ω_{DG} are the sets of buses, where the EESSs DG units are connected, respectively.

Further constrains refer to the limits imposed to the bus voltage and line current magnitudes:

$$V_{min} \leq V_{i,k_{(y,d)}} \leq V_{max} \qquad (30)$$

$$k = 1,\ldots, nt, \quad y = 1,\ldots, ny, \quad d = 1,\ldots, nd_y$$

$$I_{l,k_{(y,d)}} \leq I_{l,max} \qquad (31)$$

$$l \in \Omega_l, \quad k = 1,\ldots, nt, \quad y = 1,\ldots, ny, \quad d = 1,\ldots, nd_y$$

being V_{min} and V_{max} the minimum and maximum voltage magnitude values, $I_{l,k_{(y,d)}}$ the current flowing through the *l*th line of the set Ω_l of network lines, which cannot exceed the maximum value, $I_{l,max}$ (i.e., line ampacity).

In this case, the 2nd stage does not converge due to constraint violation, it will be iteratively applied by reducing the EESS contribution (i.e., by reducing $P_{b,i,k_{(y,d)}}$, $k = 1,\ldots, nt$), till convergence is reached.

4.4. Cost of Voltage Dips

The cost of the voltage dips is strictly connected to the economic value the effects of the voltage dips have on the equipment's and the operating processes or the activities. The most critical effect of the voltage dips is the trip of the device that is subject to the dip. The economic value of this detrimental effect represents, in turn, the cost of the voltage dip, and depends on the function of the device into the process or activity, the type of the process or activity, the linkage of the process in case stopped with other processes of the same production line, in case of industrial manufacturing loads, or with other operative functions, in case of loads different from manufacturing industries.

Generally, the cost of the voltage dips is the sum of three main components: direct, indirect and hidden costs [23]. The direct costs relate to the interruptions of the specific device or equipment. They include, for example, lost work, lost production, damaged equipment; the indirect costs include the investment costs sustained by the end user to prevent or to solve the damages due the interruptions caused by voltage dips. Finally, the hidden costs account for any second level effects that reflect on the performance of the business, such as retaining customers, satisfying customers, and protecting the company's reputation. The hidden costs in some studies are included inside the indirect costs [26].

The cost of the voltage dips can be estimated by means of direct methods or indirect methods.

The direct methods are analytical methods that require at least the availability of the voltage dips measured at the load site, the deep knowledge of several characteristics of the devices or equipment (dip susceptibility curve, operating mode, functional linkage of each device with other devices of the same equipment, and of the same process, structure and topology of the electric lines feeding each device and equipment), and the detection of the areas of the process exposed to the stoppage. They are usually conducted as after-the-fact-case studies of the events of voltage dips occurred at a specific site [26–28]. A direct method can also use the voltage dips expected at the site feeding the sensitive loads obtained by means of the simulation of the electric system in short circuit conditions. The two main methods available and widely used are the fault position method (FPM) and the critical distance method (CDM). The most adequate method to use depends on several parameters, as reported in the comparative study of [8]. Whichever of these methods obviously requires the knowledge of the electrical system feeds the load sites.

The indirect methods estimate the cost of a disturbance without passing through the analytical quantification of the effects of that disturbance on specific devices or equipment's. They can use macroeconomic analysis [11] or surveys of the customers asked to directly estimate their suffered costs due to the supply disturbances or to provide qualitative economic figures of the electric supply linked to that disturbance. In the case of the supply interruptions, for example, the National economic regulation of continuity in Italy started with an extensive survey aimed to estimate two indices: the willingness to pay (WTP) and the willingness to accept (WTA) [29]. WTP measures the additional price of the electric service the users would pay for avoiding the interruptions; WTA instead measures the amount of money they would accept for having to experience the outage. These indices allowed fixing the incentive rate for the penalties and the rewards in the first regulatory scheme of the National Authority of electricity ARERA, formerly AEEG [30].

The existence of the economic regulation of a disturbance could solve the problem of establishing the costs for the loads, especially in the planning activities with long-term horizons, as in the case of this study. In such cases, in fact, the specific details of the loads are not obtainable and would not be reliable to assign them.

Unfortunately, no economic regulation of the voltage dips is still active [31]. However, some preliminary studies [32,33] offer valuable figures resulting from measurements, survey and simulations, which can guide in deriving the economic value to assign to the voltage dips.

In this paper, we referred to [32] where the authors analyzed the economic and technical data collected for a number of medium-voltage (MV) industrial users that participated in the cost assessment project and the monitoring campaign. In particular, the authors used an indirect method based on a monitoring-and-survey approach to estimate the direct costs faced by a large variety of industrial customers. As detailed in the paper, the costs were presented both as annual value per kW and as event value for kW, with the latter independent from frequency. In this paper, we used the event value per kW of the installed power of the load.

For the evaluation of the total costs of the voltage dips over a long-time horizon, the frequency of occurrence of the voltage dips must be evaluated. The annual frequency can be estimated in average from data measured on real systems. In this paper, we referred to the measured data on the MV Italian National Grid (with reference to year 2016) at the MV busbars of the HV/MV substations (Table 1) [34].

Finally, the cost related to voltage dips over the planning period, C_{vd}, can be evaluated as:

$$C_{vd} = \sum_{i=1}^{n_{vd}} \sum_{y=1}^{n_y} \frac{1}{(1+a)^{y-1}} \sum_{rv \in \Omega_{rv}} n_{vd_{rv,y}} \, C_{vd_{rv,y}} \, P_{i,y} \qquad (32)$$

where n_{vd} is the number of busses sensitive to the voltage dips, n_y is the planning time horizon (years), a is the discount rate and, with reference to the voltage dip belonging to the residual voltage class, rv, $C_{vd_{rv,y}}$ is the cost suffered by the customer due to the single event of voltage dip at the year y, $n_{vd_{rv,y}}$ is the number of the voltage dip events at year y, and Ω_{rv} is the set of residual voltage classes which the voltage dip event refers to (i.e., first column of Table 1). $P_{i,y}$ is the power value to be considered for the dip cost identification, with reference to the user connected to the ith bus of the µG at year y. In this paper, we referred to the load nominal value.

Table 1. Annual Average Voltage Dip Number for MV substations connected to the Italian National Grid (data taken from [34]).

Residual Voltage [%]	Annual Average Voltage Dip Number				
	Duration of the Voltage Dips [ms]				
	20–200	200–500	0.5–1 × 10³	1–5 × 10³	5–60 × 10³
$80 \leq u \leq 90$	33.93	4.35	0.93	0.34	0.05
$70 \leq u \leq 80$	12.91	3.01	0.38	0.21	0.07
$40 \leq u \leq 70$	17.07	3.95	0.31	0.11	0.03
$5 \leq u \leq 40$	5.22	1.39	0.12	0.02	0.00
$1 \leq u \leq 5$	0.27	0.05	0.07	0.03	0.10
Total	69.4	12.74	1.82	0.72	0.25

We assumed that the EEC units compensate only to voltage dips at the node where the EES is installed. This assumption is very cautious, since the beneficial effects of the compensating action can be wider, involving also further nodes electrically close to the installation node of the EES. With structures of the network other from radial, the detection of the area, in which the voltage dips are compensated by the EES installed in one node, requires the simulation of the system in short circuit conditions with EES acting as voltage dip compensating unit. A FPM, applied in the presence of the EES, giving as result the propagation of the voltage dips for short circuits in every node of the system, could allow obtaining this result.

5. Numerical Applications

The planning procedure has been applied to an MV µG, which refers to the MV Cigrè benchmark system [35]. The system under study is a 12.47 kV, three-phase, balanced, distribution network (Figure 3). This network is constituted by 15 buses and it is connected to an upstream HV network by means of two 115/12.47 kV transformers of 25 and 20 MVA, which connect the PCC (bus #1) to two feeders: the first feeder is connected at the secondary side of the 25 MVA transformer (bus #2); the second feeder is connected at the secondary side of the 20 MVA transformer (bus #13). The network includes three switches: S_1 (between buses #9 and #15), S_2 (between buses #7 and #8) and S_3 (between buses #5 and #12).

The loads connected to the network are grouped in residential and commercial/industrial and are listed in Table 2. The loads are assumed to have a yearly growth of 1%. The per unit power profiles assumed for residential and industrial/commercial customers are reported in Figure 4. Two PV units are connected to the buses #9 and #15 with rated powers of 5 MW and 2 MW, respectively. Figure 5 shows the per unit power profile of the PV generation at bus #9.

The assumed energy pricing tariff refers to the hourly values of the Italian market price. With reference to the first year, the price values refer to a day of November 2019 [36] (Figure 6). A yearly growth of 1% has also been assumed for the energy prices. Regarding the cost, which the customers have to sustain due to the single voltage dip event, the value 2.9 €/kW [32] has been assumed with a yearly growth of 1% for the industrial/commercial loads; for the residential loads, instead, the value of zero was assumed. The number of events is evaluated according to the data reported in Table 1. Particularly, the events related to the threshold voltage dip value, V_{cr}, of 70% has been assumed.

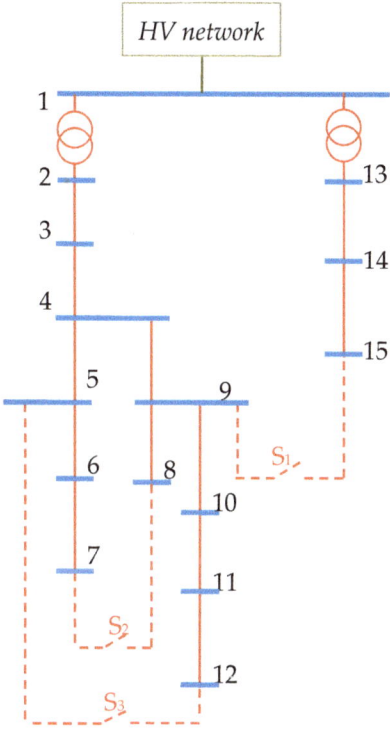

Figure 3. The microgrid under study.

Table 2. Load Input data.

Bus #	Rated Power (MVA)	Load Type	cos φ	Bus #	Rated Power (MVA)	Load Type	cos φ
2	13.8	Residential	0.93	8	0.30	Comm./industrial	0.95
	9.16	Comm./industrial	0.87	9	0.25	Residential	0.90
3	0.35	Residential	0.95		0.20	Comm./industrial	0.90
	0.80	Comm./industrial	0.85	10	0.35	Residential	0.95
4	0.25	Residential	0.90	11	0.50	Residential	0.90
	0.24	Comm./industrial	0.80	12	0.10	Residential	0.95
5	0.40	Residential	0.90		0.45	Comm./industrial	0.85
6	0.20	Residential	0.95	13	3.20	Residential	0.90
	0.30	Comm./industrial	0.85		3.78	Comm./industrial	0.87
7	0.15	Residential	0.95	14	0.68	Comm./industrial	0.85
8	0.10	Residential	0.95	15	0.27	Comm./industrial	0.90

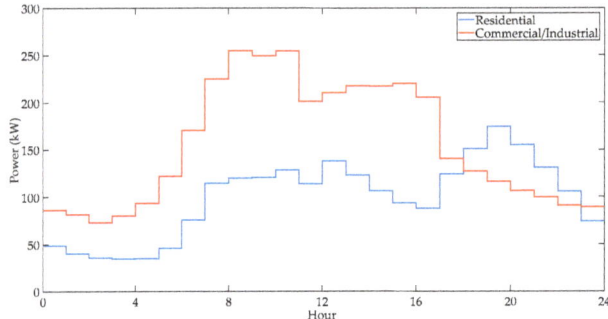

Figure 4. Load profile at bus #6.

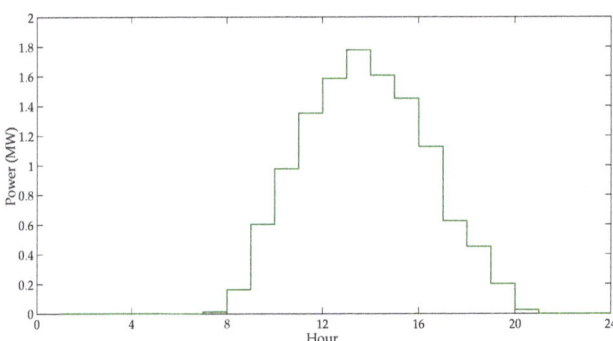

Figure 5. PV power profile at bus #15.

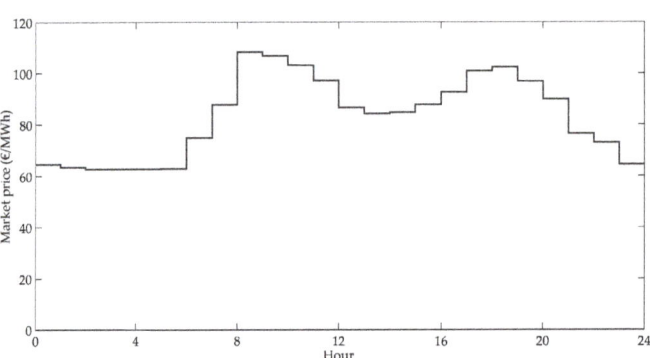

Figure 6. Market energy price (data taken from [36]).

Regarding the EESSs to install, nickel manganese cobalt Li-ion battery technology has been considered suitable for this application. The data of the various EESS' features are reported in Table 3 [37], which refer to values expected for 2020. Regarding the battery replacement cost, the value of the cost expected for the 2025 has been considered. Note that the costs in Table 3 have been converted in Euro (conversion rate of November 2019) before applying them in the proposed procedure.

Table 3. EESS Data (data taken from [37]).

Parameter	Value	Parameter	Value
Battery round trip efficiency	99 %	Battery Cycle life	4800 cycles
Maximum Depth of Discharge	100 %	Converter round trip efficiency	98 %
Battery installation cost	153 $/kWh	Converter installation cost	59.6 $/kW
Battery replacement cost	110 $/kWh	Maintenance cost	1.5%

The planning design alternatives refer to the installation of two EESSs whose size must be selected among the rated power values 0, 5, 10 and 15 MW with a discharging nominal time of five hours.

In order to show the ability of the method to satisfy the technical constraints imposed on the EESSs and on the µG, Figures 7 and 8 report the daily profiles of the active power of an EESS and of the voltage values at the buses where an EESS and a PV unit are connected. The figures refer to the first year of the planning period of the case study 1 and detail the profiles of the EESS connected at the bus #6 and of the PV unit connected at the bus #15.

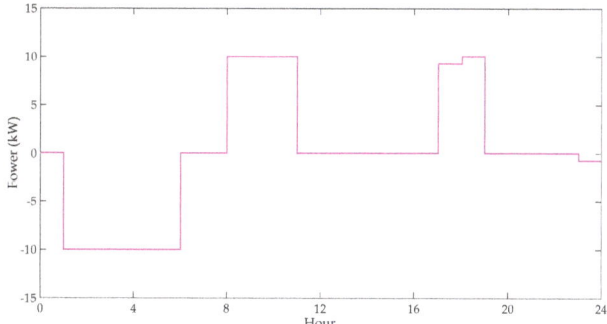

Figure 7. Active power of the EESS connected at the bus #6 (case study 1).

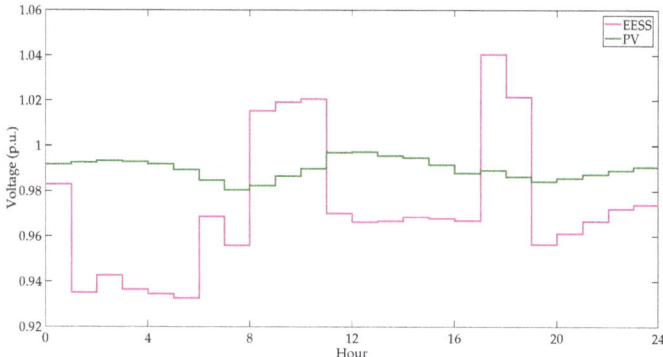

Figure 8. Voltage values along a day of the first year of the planning period at the bus where EESS (bus #6) and PV unit (bus #16) are connected (case study 1).

During the µG operation, the breakers can be operated open or closed depending on the operator decisions aimed at avoiding undesired conditions or based on economic strategies. The states of the breakers strongly affect the values that line currents and bus voltages can assume. In order to consider the influence of the states of the breakers in terms of exposure of the buses to the voltage dips, the proposed planning procedure has been applied with reference to the following case studies:

- Case 1: S_1, S_2, S_3: open
- Case 2: S_1, S_2, and S_3: closed
- Case 3: S_1: closed, S_2, S_3: open
- Case 4: S_1, S_3: open, S_2: closed
- Case 5: S_1, S_2: open, S_3: closed
- Case 6: Case 1 (0.45), Case 2 (0.10), Case 3 (0.15), Case 4 (0.15), Case 5 (0.15)

Case 6 refers to the probability of occurrence of different network configurations (the probability of occurrence of each configuration is indicated within the brackets).

The application of the procedure of Section 3 for the identification of candidate buses based on voltage dip sensitiveness leads to the candidate buses reported in Table 4 that correspond to the threshold voltage dip value of 70%. The results of the planning procedure corresponding to each case study are reported in Table 5. In the table, the BF of each planning solution is reported together with the reduction of the cost of voltage dips obtained thanks to the installation of the EESSs. For comparative purposes, the planning results are reported for both the cases in which the reduction of voltage dip cost is taken into account and those in which this cost reduction is neglected. Negative values of BF are obtained in the cases where installation of EESSs is not convenient.

Table 4. Candidate buses (70% threshold values).

Case Study					
Case 1	Case 2	Case 3	Case 4	Case 5	Case 6
3	6	8	3	3	8
4	7	9	4	4	9
5	8	10	5	5	10
6	9	11	6	6	11

Table 5. Siting and sizing of EESSs.

Case Study	Accounting for Voltage Dip Costs	BF (M$)	Voltage Dip Cost Reduction (M$)	EESSs' Nodes		EESSs' Sizes (MW)	
				EESS n. 1	EESS n. 2	EESS n. 1	EESS n. 2
1	Yes	4.81	1.63	3	6	10	10
	No	3.19	-	3	6	10	10
2	Yes	−0.16	0.44	6	-	5	-
	No	−0.59	-	9	-	5	-
3	Yes	−0.34	0.31	9	-	5	-
	No	−0.65	-	9	-	5	-
4	Yes	4.75	1.63	3	6	10	10
	No	3.16	-	3	4	15	5
5	Yes	4.70	1.63	3	6	10	10
	No	3.21	-	3	5	5	15
6	Yes	4.46	0.50	8	10	15	10
	No	3.96	-	8	10	15	10

The results reported in Table 5 show that the convenience of installing EESSs in the µG basically depends on the cost reduction obtained thanks to the price arbitrage. In fact, in the cases where the price arbitrage does not allow installing EESSs (cases 2 and 3), the added benefits derived from the voltage dip compensation is not sufficient to justify the adoption of the storage devices. This is probably due to the fact that, being the candidate buses mainly characterized by residential loads, the benefit derived by dip compensation in terms of cost reduction is not significant (we remind that the cost of voltage dip for the residential load was neglected). Moreover, in the cases where it is convenient to install the EESSs (cases 1, 4, 5, and 6) the beneficial effect of dip compensation clearly appears only when they are used in buses with large industrial loads. In fact, in cases 4 and 5, the installation of the EESS n. 2 in bus #6 instead of bus #4 and in bus #6 instead of bus #5, respectively, allows obtaining higher benefits.

In order to better prove the effect of the dip compensation on the total cost reduction, in Table 6, the percentage values of the benefit related to the voltage dip cost reduction are reported, with reference to the cases in which the planning procedure provides a solution. It is interesting to note that the percentage benefit due to the voltage dip cost reduction falls within the range [33–35] % in cases 1, 4, and 5. In case 6, the percentage value is lower (about 11%). This is mainly due to the fact that in case 6, the solution associated to the largest total benefit corresponds to allocation buses (buses #8 and #10) with voltage dip costs lower than those corresponding to the nodes selected in cases 1, 4, and 5.

Table 6. Benefit corresponding to the voltage dip cost reductions.

Case Study	Percentage Value of the Voltage Dip Cost Reduction on the Total Benefit (%)	Percentage of Value of the Voltage Dip Cost Reduction on the Installation and Replacement Cost (%)
1	33.89	6.55
4	34.32	6.55
5	34.68	6.55
6	11.21	1.61

In Table 6, the percentage values of the benefit corresponding to the voltage dip reduction compared to the EESS's installation and replacement costs are also reported. The value of about 6% is obtained in the cases 1, 4 and 5. A lower value (about 2%), instead, is obtained in case 6. This is due to two aspects: the first refers again to the reduced benefit in terms of voltage dip cost reduction (compared to the other cases), the second refers to the increased size of storage, which is allocated in this case study.

Figure 7 clearly shows that the maximum power that is charged/discharged by the EESS is always lower than the rated power. Moreover, the comparison of Figures 6 and 7, reveals that the battery absorbs power during the lowest price hours, and injects power to the grid during the highest price hours. This is coherent with the considered minimum cost strategy. The SoC profile—here not reported for the sake of conciseness—also satisfies the constraints on the rated battery capacity.

Figure 8 shows that the voltage at the critical buses, i.e., the buses where an EESS or a DG are connected, falls within admissible ranges (0.9–1.1) p.u. The increased variation of the voltage in the nodes where an EESS is connected which appears in the figure is due to the charge/discharge of the EESS. In the node where a DG is connected, instead, a smooth profile appears; this is due to the power rating of the generation system lower than that of the EESS and to the presence of the high load power in the buses close to this node.

Still referring to the case study 1—first year of the planning period—Figure 9 shows the active power imported at the PCC. Coherently with the minimum cost strategy, thanks to the use of the EESSs, most of the requested power is shifted from the highest price hours to the lowest price hours.

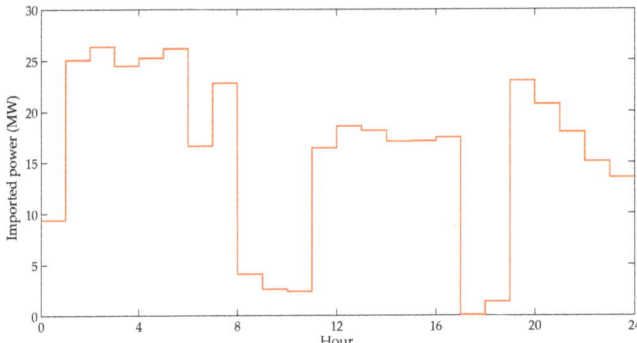

Figure 9. Active power imported from the whole µG along a day of the first year of the planning period.

6. Conclusions

In this paper, the planning of the electrical energy storage systems in the microgrids was considered. A cost minimization approach was proposed including the benefits obtained through the voltage dip compensation. Particularly, the cost reduction is obtained by shifting the load from the hours of high energy price to those of low energy prices and by considering the voltage dip cost avoided thanks to the compensation action. The motivation of this study is twofold. The first relies on the importance of power quality aspects in the modern power systems characterized by widespread use of electronic devices and control systems which are highly vulnerable to the voltage dips. The latter refers to the benefits derived from the ability the storage systems have in providing multiple services, which concur to sustain their high installation cost. The proposed method is based on a multi-step planning tool which allows identifying the optimal location and size of the storage systems. The results of the numerical simulations clearly showed that voltage dip compensation allowed obtaining non-negligible economic benefits. Clearly, these benefits strictly depend on the cost assumed for the voltage dip events, which can be specified only through a deep knowledge of the end-users. Future works will be aimed at improving the procedure in order to better identify the areas of the network that achieve a beneficial effect in terms of dip compensation from the installation of EESSs.

Author Contributions: F.M., D.P., P.V. (Pietro Varilone) and P.V. (Paola Verde) conceived and designed the theoretical methodology and the numerical applications; F.M., D.P., P.V. (Pietro Varilone) and P.V. (Paola Verde) performed the numerical applications and analyzed the results; F.M., D.P., P.V. (Pietro Varilone) and P.V. (Paola Verde) wrote and revised the paper. All authors have read and agreed to the published version of the manuscript.

Funding: Financial support by Italian Ministry of University and Research (MIUR) through the Special Grant "Dipartimenti di eccellenza" Lex. N. 232/31.12.2016 (G.U. n. 297/21.12.2016 S.O. n. 57).

Acknowledgments: The authors thank Guido Carpinelli (University of Naples Federico II) for his invaluable and generous contribution in the continuous development of new ideas to deal with increasingly interesting problems of modern power systems. The authors P. Varilone and P. Verde acknowledge the financial support by Italian Ministry of University and Research through the Special Grant "Dipartimenti di eccellenza".

Conflicts of Interest: The authors declare no conflict of interest.

References

1. Ai Wong, L.; Ramachandaramurthy, V.K.; Taylor, P.; Ekanayake, J.B.; Walker, S.L.; Padmanaban, S. Review on the optimal placement, sizing and control of an energy storage system in the distribution network. *J. Energy Storage* **2019**, *21*, 489–504. [CrossRef]
2. Hemmati, R.; Shafie-Khah, M.; Catalão, J.P.S. Three-Level Hybrid Energy Storage Planning Under Uncertainty. *IEEE Trans. Ind. Electron.* **2019**, *66*, 2174–2184. [CrossRef]

3. Zhang, N.; Li, R.; Jiang, Y. Cost-benefits analysis of battery storage system for industry consumers based on different operation modes. In Proceedings of the 2018 IEEE 2nd International Electrical and Energy Conference (CIEEC), Beijing, China, 4–6 November 2018; pp. 527–531. [CrossRef]
4. Saboori, H.; Hemmati, R.; Sadegh Ghiasi, S.M.; Dehghan, S. Energy storage planning in electric power distribution networks—A state-of-the-art review. *Renew. Sustain. Energy Rev.* **2017**, *79*, 1108–1121. [CrossRef]
5. Carpinelli, G.; Celli, G.; Mocci, S.; Mottola, F.; Pilo, F.; Proto, D. Optimal Integration of Distributed Energy Storage Devices in Smart Grids. *IEEE Trans. Smart Grid* **2013**, *4*, 985–995. [CrossRef]
6. Grover-Silva, E.; Girard, R.; Kariniotakis, G. Optimal sizing and placement of distribution grid connected battery systems through an SOCP optimal power flow algorithm. *Appl. Energy* **2018**, *219*, 385–393. [CrossRef]
7. Carpinelli, G.; Mottola, F.; Proto, D. Probabilistic sizing of battery energy storage when time-of-use pricing is applied. *Electr. Power Syst. Res.* **2016**, *141*, 73–83. [CrossRef]
8. Electricity Energy Storage Technology Options: A White Paper Primer on Applications Costs and Benefits. 2010. Available online: https://www.epri.com/ (accessed on 12 January 2020).
9. Carpinelli, G.; Di Perna, C.; Caramia, P.; Varilone, P.; Verde, P. Methods for Assessing the Robustness of Electrical Power Systems Against Voltage Dips. *IEEE Trans. Power Deliv.* **2009**, *24*, 43–51. [CrossRef]
10. Caramia, P.; Varilone, P.; Verde, P.; Vitale, L. Tools for Assessing the Robustness of Electrical System against Voltage Dips in terms of Amplitude, Duration and Frequency. In Proceedings of the International Conference on Renewable Energies and Power Quality (ICREPQ'14), Cordoba, Spain, 8–10 April 2014.
11. Küfeoğlu, S.; Lehtonen, M. Macroeconomic Assessment of Voltage Sags. *Sustainability* **2016**, *8*, 1304. [CrossRef]
12. Ding, Z.; Zhu, Y.; Chen, C. Economic loss assessment of voltage sags. In Proceedings of the CICED 2010 Proceedings, Nanjing, China, 13–16 September 2010; pp. 1–5.
13. *6th CEER Benchmarking Report on the Quality of Electricity and Gas Supply*; CEER, 2016. Available online: https://www.ceer.eu/1305/ (accessed on 12 January 2020).
14. Nick, M.; Cherkaoui, R.; Paolone, M. Optimal siting and sizing of distributed energy storage systems via alternating direction method of multipliers. *Int. J. Electr. Power Energy Syst.* **2015**, *72*, 33–39. [CrossRef]
15. Nojavan, S.; Majidi, M.; Esfetanaj, N.N. An efficient cost-reliability optimization model for optimal siting and sizing of energy storage system in a microgrid in the presence of responsible load management. *Energy* **2017**, *139*, 89–97. [CrossRef]
16. Lin, Z.; Hu, Z.; Zhang, H.; Song, Y. Optimal ESS allocation in distribution network using accelerated generalised Benders decomposition. *IET Gener. Transm. Distrib.* **2019**, *13*, 2738–2746. [CrossRef]
17. Fantauzzi, M.; Lauria, D.; Mottola, F.; Scalfati, A. Sizing energy storage systems in DC networks: A general methodology based upon power losses minimization. *Appl. Energy* **2017**, *187*, 862–872. [CrossRef]
18. Nick, M.; Cherkaoui, R.; Paolone, M. Optimal Allocation of Dispersed Energy Storage Systems in Active Distribution Networks for Energy Balance and Grid Support. *IEEE Trans. Power Syst.* **2014**, *29*, 2300–2310. [CrossRef]
19. Andreotti, A.; Carpinelli, G.; Mottola, F.; Proto, D.; Russo, A. Decision Theory Criteria for the Planning of Distributed Energy Storage Systems in the Presence of Uncertainties. *IEEE Access* **2018**, *6*, 62136–62151. [CrossRef]
20. Celli, G.; Pilo, F.; Pisano, G.; Soma, G.G. Including voltage dips mitigation in cost-benefit analysis of storages. In Proceedings of the 2018 18th International Conference on Harmonics and Quality of Power (ICHQP), Ljubljana, Slovenia, 13–16 May 2018; pp. 1–6. [CrossRef]
21. IEC 61000-4-11:2004. *Testing and Measurement Techniques—Voltage Sags, Short Interruptions and Voltage Variations Immunity Tests*. 2004. Available online: https://webstore.iec.ch (accessed on 12 January 2020).
22. Carpinelli, G.; Caramia, P.; Di Perna, C.; Varilone, P.; Verde, P. Complete Matrix Formulation of Fault Position Method for Voltage Dip Characterization. *IET Gener. Transm. Distrib.* **2007**, *1*, 56–64. [CrossRef]
23. IEEE Standard 1346. *Recommended Practice for Evaluating Electric Power System Compatibility with Electronic Process Equipment*; The Institute of Electrical and Electronics Engineers, Inc.: New York, NY, USA, 1998.
24. Xu, B.; Oudalov, A.; Ulbig, A.; Andersson, G.; Kirschen, D.S. Modeling of Lithium-Ion Battery Degradation for Cell Life Assessment. *Trans. Smart Grid* **2018**, *9*, 1131–1140. [CrossRef]
25. Han, X.; Lu, L.; Zheng, Y.; Feng, X.; Li, Z.; Li, J.; Ouyang, M. A review on the key issues of the lithium ion battery degradation among the whole life cycle. *ETransportation* **2019**, *1*, 100005. [CrossRef]

26. Fumagalli, E.; Lo Schiavo, L.; Delestre, F. *Service Quality Regulation in Electricity Distribution and Retail*; Springer: Berlin/Heidelberg, Germany, 2007.
27. Di Fazio, A.R.; Duraccio, V.; Varilone, P.; Verde, P. Voltage sags in the automotive industry: Analysis and solutions. *Electr. Power Syst. Res.* **2014**, *110*, 25–30. [CrossRef]
28. Weldemariam, L.; Cuk, V.; Cobben, J. Cost Estimation of Voltage Dips in Small Industries Based on Equipment Sensitivity Analysis. *Smart Grid Renew. Energy* **2016**, *7*, 271–292. [CrossRef]
29. Bertazzi, A.; Fumagalli, E.; Lo Schiavo, L. The use of customer outage cost surveys in policy decision-making: The Italian experience in regulating quality of electricity supply. In Proceedings of the CIRED 2005—18th International Conference and Exhibition on Electricity Distribution, Turin, Italy, 6–9 June 2005; pp. 1–5.
30. Delibera n. 155/02, *Testo Integrato delle Disposizioni Dell'autorità per L'energia Elettrica e il Gas in Materia di Continuità del Servizio di Distribuzione Dell'energia Elettrica, G.U. Serie Generale n. 201 del 28 Agosto 2002.* Available online: https://www.arera.it/it/docs/02/155-02.htm (accessed on 12 January 2020). (In Italian)
31. Bollen, M.; Beyer, Y.; Styvactakis, E.; Trhulj, J.; Vailati, R.; Friedl, W. A European Benchmarking of voltage quality regulation. In Proceedings of the 2012 IEEE 15th International Conference on Harmonics and Quality of Power, Hong Kong, China, 17–20 June 2012; pp. 45–52.
32. Delfanti, M.; Fumagalli, E.; Garrone, P.; Grilli, L.; Lo Schiavo, L. Toward Voltage-Quality Regulation in Italy. *IEEE Trans. Power Deliv.* **2010**, *25*, 1124–1132. [CrossRef]
33. Weldemariam, L.E.; Cuk, V.; Cobben, J.F.G. A proposal on voltage dip regulation for the Dutch MV distribution networks. *Int. Trans Electr. Energy Syst.* **2019**, *29*, e2734. [CrossRef]
34. *Relazione Annuale Sullo Stato dei Servizi e Sull' Attività Svolta*. AEEGSI Annual Report AEEGSI. 2018. Available online: www.autorita.energia.it (accessed on 12 January 2020). (In Italian)
35. *Benchmark Systems for Network Integration of Renewable and Distributed Energy Resources, Cigré Task Force C6.04, Cigré Brochure 575.* 2014. Available online: https://e-cigre.org/ (accessed on 12 January 2020).
36. Gestore Mercati Energetici. Esiti Mercato Elettrico, November 2019. Available online: https://www.mercatoelettrico.org/it/ (accessed on 12 January 2020).
37. IRENA. Electricity Storage and Renewables: Costs and Markets to 2030, International Renewable Energy Agency, Abu Dhabi. 2017. Available online: http://www.irena.org/publications/2017/Oct/Electricity-storage-and-renewables-costs-and-markets (accessed on 12 January 2020).

© 2020 by the authors. Licensee MDPI, Basel, Switzerland. This article is an open access article distributed under the terms and conditions of the Creative Commons Attribution (CC BY) license (http://creativecommons.org/licenses/by/4.0/).

Article

Optimal Energy Storage System Positioning and Sizing with Robust Optimization

Nayeem Chowdhury, Fabrizio Pilo * and Giuditta Pisano

Department of Electrical and Electronic Engineering, University of Cagliari, 09123 Cagliari, Italy; nayeem.chowdhury@enel.com (N.C.); giuditta.pisano@unica.it (G.P.)
* Correspondence: fabrizio.pilo@unica.it; Tel.: +39-320-437-2957

Received: 2 December 2019; Accepted: 17 January 2020; Published: 21 January 2020

Abstract: Energy storage systems can improve the uncertainty and variability related to renewable energy sources such as wind and solar create in power systems. Aside from applications such as frequency regulation, time-based arbitrage, or the provision of the reserve, where the placement of storage devices is not particularly significant, distributed storage could also be used to improve congestions in the distribution networks. In such cases, the optimal placement of this distributed storage is vital for making a cost-effective investment. Furthermore, the now reached massive spread of distributed renewable energy resources in distribution systems, intrinsically uncertain and non-programmable, together with the new trends in the electric demand, often unpredictable, require a paradigm change in grid planning for properly lead with the uncertainty sources and the distribution system operators (DSO) should learn to support such change. This paper considers the DSO perspective by proposing a methodology for energy storage placement in the distribution networks in which robust optimization accommodates system uncertainty. The proposed method calls for the use of a multi-period convex AC-optimal power flow (AC-OPF), ensuring a reliable planning solution. Wind, photovoltaic (PV), and load uncertainties are modeled as symmetric and bounded variables with the flexibility to modulate the robustness of the model. A case study based on real distribution network information allows the illustration and discussion of the properties of the model. An important observation is that the method enables the system operator to integrate energy storage devices by fine-tuning the level of robustness it willing to consider, and that is incremental with the level of protection. However, the algorithm grows more complex as the system robustness increases and, thus, it requires higher computational effort.

Keywords: decision-making; distribution network planning; uncertainty; robust optimization; energy storage system

1. Introduction

The share of renewable power generation in the global electricity generation is anticipated to expand from today's 23% to levels between 30%–45% by 2030 [1]. This technological alteration requires a rethinking in the way power systems are planned, maximize the benefits from renewables affordably and securely. Since renewable energy integration brings new challenges into the distribution network planning an accurate planning model, which incorporates system uncertainty introduced by renewable resources and loads, is necessary for making planning decision.

The technological development of large-scale electrochemical energy storage system (ESS) has resulted in capital cost reductions and increased roundtrip efficiency enables them to become a feasible option to deploy in the distribution network [2,3]. Storage applications such as energy arbitrage [4], peak shaving [5], frequency regulation [6], voltage support [7], and congestion management [8] have made it vital to integrate more ESS in the distribution network. Thus, optimal planning and management of ESS are essential to identify ideal configurations. However, many of the optimization algorithms proposed in recent literature do not adequately deal with uncertainties. For instance, the real amount and position of distributed generation (DG) that is going to be connected to the system, the mix of renewable energy sources (RES), the cost of ESS or the level of participation and the cost for active demand [9].

Sizing of ESSs in distribution networks from DSO has been discussed in [10]. The number and locations of the ESSs are assumed to be given. An AC-optimal power flow (AC-OPF) with semidefinite programming (SDP) convex relaxation is adopted for network simulation. To consider uncertainties in the model, a stochastic optimization approach has been considered. Two different problems have been formulated respectively for the siting and sizing of ESS in distribution networks coupled with a wind farm in [11]. However, the authors considered a linearized DC-optimal power flow (DC-OPF), and the wind power forecast is assumed perfect. Apart from siting and operation, the authors of [12] suggest the life cycle payment of storage. They present two models for a transmission-constrained power network with storage. Both models use a DC-OPF framework. The first model selects optimal siting and operation of the storage assuming a fixed group of different storage technologies. The second model expands the DC-OPF framework to optimize the storage technology mix, new storage capacity investments, and the network allocation of these resources. The authors of [13] provide a mathematical model that simultaneously optimizes transmission switching operations, ESS siting and sizing decisions and taking into account the limits on maximum allowable load shedding and renewable energy curtailment amounts in the power system. The methodology proposed in [4], based on a linearized DC-OPF, captured both the monetary and technical advantages of investment in storage and adopted a sensitivity analysis to assess the impact of uncertain parameters. As opposed to an analytical approach, the authors of [14] detail a heuristic approach for finding the optimal location(s) and size of a multi-purpose ESS including transmission and distribution parts without considering the uncertainty in the model. In the transmission storage part, a sensitivity analysis is performed using complex-valued neural networks (CVNN) and time domain power flow (TDPF) to obtain the optimal ESS location(s). In [15] a multi-criteria approach where a genetic algorithm (NSGA-II) has been used to identify the optimal place, size, and scheduling of energy storage in the distribution network. The authors created a full multi-objective (MO) optimization procedure able to identify the Pareto set of design options with fixed network topology for a given medium voltage (MV) network. In addition to that, the same authors of [15] have proposed a multi-criteria analysis approach selecting the best planning alternative for energy storage integration in the distribution system [16]. However, heuristic techniques often required a high computational burden and are not guaranteed to converge in global optima [17].

Convex relaxation techniques have been developed to obtain an acceptable solution while ensuring algorithmic efficiency. The two most commonly used relaxations for distribution network are semi-definite program (SDP) and second-order cone programming (SOCP). Though both SOCP and SDP have been proven exact under certain conditions [18,19]. In this paper, SOCP has been adopted due to its higher algorithmic performances that imply fast convergence to global optima and to reduce the heavy computation cost.

The current literature on energy storage study is divided into three classifications: (i) storage sizing, (ii) storage operation, and (iii) storage siting. Less publications exist about the optimal location(s) of the ESS than publications on optimal sizing likely due to the difficulty of finding optimal sites [20,21]. Storage siting is the least researched and most complicated of these three classifications. The optimal operation studies of ESS consider that energy and power ratings of a storage unit are given, the purpose of these studies is to identify operation strategies to optimize the exploitation of resources able of contributing to network support at minimum cost. These studies typically do not address network constraints. The optimal ESS size (i.e., energy and power ratings) depends on the state at which the storage is optimally operated [22]. In turn, optimal storage siting depends on the amount of the storage being considered and how it will be controlled. This problem becomes even more complicated when it considers distributed storage rather than a single storage unit. If this is the case, the amount of storage located is typically undefined at first. ESSs are considered one of the solutions to manage the DG downsides and to help incorporate RES in the distribution networks, which rely largely on the flexibility of resources [23,24].

Capital intensity is the main barrier to the deployment of ESS [25]. Investment in the ESS, therefore, requires a trade-off between long-term investment costs, short-term operating conditions, and the benefits that these services will offer. It includes optimizing their network position (i.e., location) and operating parameters (e.g., size and operating profile) simultaneously [26]. The objective of the optimization is generally multiple since it spans from the reduction of CAPEX (Capital Expenditures) for network upgrade to the reduction of energy loss costs as well as power quality costs [27]. RES and electric vehicles (EVs) integration as well as the engagement of customers in flexibility programs are opening new opportunities for ESS and making clear the need for optimal siting and sizing methodologies [28].

ESS technologies can operate on different timescales, ranging from seconds to hours. The services offered by ESS can be divided into power- and energy-related services, based on the timescale of interest [29]. Transient stability and ancillary services, such as frequency regulation, spinning reserve, and voltage control are power-related services. Back-up power provision, black-start, uninterruptible power supply (UPS), standing reserve, and seasonal energy storage are typical examples of energy-related services [28,30]. Both ESS owners and other system stakeholders can benefit from the provision or the usage of these services.

The ESS optimal positioning and sizing problem aims at the maximization of the benefit-cost ratio subject to the non-linear/non-convex network constraints that make the solution more cumbersome and requires specific mathematical tool [31]. The research on the topic has been dramatically increasing in the last two years since ESS are crucial for the energy transition towards the carbon free world, but there is still room for new contributions, particularly on dealing with the uncertainties modeling.

For these reasons, the paper proposes an application of the Robust Optimization (RO) to solve the ESS optimal location problem in distribution networks operated by a DSO. The objective of the optimization problem is to use ESS for delivering power without any violation of technical limits (e.g., maximum and minimum nodal voltage), minimizing the resort to RES or combined heat and power (CHP) generation curtailment and load shaving. Indeed, the DSO can evaluate the ESS installation as a non-network option to avoid the expensive and time consuming building or revamping of networks, particularly in the current situation of limited markets of services offered by customers or other producers. However, the convenience to do this strictly depends on the site, size, and operation of the ESS, that in turn, depend on the state of the network, that is intrinsically uncertain. The approach proposed in the paper deals with the uncertainties with an original implementation of the Robust Optimization. The algorithm has been validated with an exemplary distribution network representative of one class of the Italian distribution classes of networks produced by the project ATLANTIDE (Archivio TeLemAtico per il riferimento Nazionale di reTI di Distribuzione Elettrica" that means "Digital archive for the national electrical distribution reference networks") [32].

The paper is organized as follows: Section 2 describes the detailed formulation of energy storage placement problem. Section 3 discusses the uncertainty modeling approach. Section 4 describes the solution methodology of a robust optimization problem. Sections 5 and 6 present the case study with real data and conclusion, respectively. Finally, in the nomenclature of the symbols used in the mathematical formulation has been reported.

2. Deterministic Formulation of Energy Storage Planning

The objective function (OF) of the deterministic model consists of minimizing the operational extra-cost that should be sustained for complying with the technical constraints. Such cost includes the penalty terms for RES (C_n^{RESc}) and biomass CHP generation curtailment (C_n^{CHPc}), and the cost of shaving the peak loads (C_n^{PLS}). Furthermore, since the goal of the paper is to evaluate the contribution of energy storages to the management of the network, even in uncertain conditions, the investment cost $C_n^{CAPEX\_ESS}$ to be sustained for the storage allocated in the network is added to the operational cost, as in (1).

$$\min C_{tot} = \min\left\{\sum_{n=1}^{N}\left[C_n^{RESc} + C_n^{CHPc} + C_n^{PLS} + C_n^{CAPEX\_ESS}\right]\right\} \quad (1)$$

This minimization is subject to voltage and current limits, power flow equations, and storage technical constraints. In the following, each cost term and constraints are detailed.

2.1. Penalty for RES Curtailment C_n^{RESc}

To strongly penalize the generation curtailment of RES, the cost of curtailed energy due to network constraint violations has been monetized as twice the price of energy paid in the wholesale market c_{EN} (here, 58 €/MWh, according to the average Italian energy selling price) [33], as in (2).

$$C_n^{RESc} = \sum_{t=1}^{T} 2 \cdot c_{EN} \cdot P_n^{RESc}(t) \quad n = 1 \cdots N \quad (2)$$

where $P_n^{RESc}(t)$ is the energy curtailed at the time interval t by the RES generator connected to the n-th bus of the network.

Since the increment of the network hosting capacity may be quantified via the possibly avoided curtailment of RES production, the smaller this term, the better the storage allocation solution.

2.2. Penalty for Biomass CHP Curtailment C_n^{CHPc}

This cost for biomass CHP curtailment is assumed proportional to the avoided cost for fuel saving F [€/MWh], increased by 20%. The fuel cost F has been considered here equal to 80 €/MWh, by assuming the average gas price ≈ 35 c€/m³ ≈ 36 €/MWh$_t$ and by hypothesizing an efficiency for the electric conversion about 45%. This assumption allows penalizing also the CHP curtailments, with high cost, as in (3).

$$C_n^{CHPc} = \sum_{t=1}^{T} 1.2 \cdot F \cdot P_n^{CHPc}(t) \quad n = 1 \cdots N \quad (3)$$

where $P_n^{CHPc}(t)$ is the energy curtailed at the time interval t by the biomass CHP connected to the n-th bus of the network.

2.3. Peak Load Shaving Cost C_n^{PLS}

Regarding the term referred to the active customers, in this paper only the cost of shaving the peak loads has been considered, by assuming that it is not possible to fully control the customer demand but only cut a quote of their consumption in some critical conditions. It is assumed, as the RES curtailment, that this curtailed energy is paid at twice the energy price c_{EN} to penalize load curtailment with the higher cost, as renewable generation curtailment, according to (4).

$$C_n^{PLS} = \sum_{t=1}^{T} 2 \cdot c_{EN} \cdot P_n^{PLS}(t) \quad n = 1 \cdots N \quad (4)$$

where $P_n^{PLS}(t)$ is the energy curtailed at the time interval t to the customer connected to the n-th bus of the network.

2.4. Storage Investment Cost $C_n^{CAPEX\_ESS}$

The storage investment cost (SC_n) is a function of the size of the storage in terms of rated power and energy as in (5).

$$SC_n = c_P \cdot P_n^{rated} + c_E \cdot E_n^{rated} \quad n = 1 \cdots N \tag{5}$$

where c_P and c_E are the specific costs of the ESS adopted technology, reliant respectively on the power rating P_n^{rated} and the nominal capacity E_n^{rated} of the n-th ESS located in the network (here c_P = 200 €/kW and c_E = 400 €/kWh, according to the market cost of lithium-ion technology [15]).

To consider this cost in the objective function (1), only a daily quote of SC_n is added to the operational terms of (1), calculated as in (6).

$$C_n^{CAPEX\_ESS} = \frac{K_S}{365} \cdot SC_n \quad n = 1 \cdots N \tag{6}$$

where K_s is a capital recovery factor (here K_s = 0.1, for considering 10 years as ESS lifetime).

In this paper, it is assumed that the storages are DSO owned and managed for relieving contingencies. Thus, the ESS OPEX (operational expenditures) is not considered in the optimization. According to this point of view, it is supposed that the minimization of the network operational cost, in terms of reduction of the curtailed power from RES and to loads, that would be necessary to relieve contingencies, represents the only incomes that allow DSO to pay back EES CAPEX (capital expenditures) and ESS OPEX. The depreciation of the ESSs is assumed negligible and not added to the ESS cost term.

2.5. Load Balancing Constraints

The SOCP convex relaxation has been used in the proposed multi-temporal AC-OPF model. Equations (7) and (8) are the nodal active and reactive power balance.

$$P_n^g(t) + P_n^{RES}(t) - P_n^{RESc}(t) + P_n^{CHP}(t) - P_n^{CHPc}(t) - PD_n(t) + P_n^{PLS}(t) - P_n^c(t) + P_n^d(t) \\ - \sum_{m \in \theta_n} R_{mn} \cdot I_{mn}^2 = \sum_{m \in \theta_n} P_{mn}(t) \tag{7}$$

$$Q_n^g(t) + Q_n^{RES}(t) - Q_n^{RESc}(t) + Q_n^{CHP}(t) - Q_n^{CHPc}(t) - QD_n(t) - \sum_{m \in \theta_n} X_{mn} \cdot I_{mn}^2 = \sum_{m \in \theta_n} Q_{mn} \tag{8}$$

where $(P_n^{RES}(t); Q_n^{RES}(t))$ and $(P_n^{CHP}(t); Q_n^{CHP}(t))$ define the expected RES and CHP production in terms of active and reactive powers, $PD_n(t)$ and $QD_n(t)$ are the active and reactive power delivered to the load connected to the n-th node, $I_{mn}(t)$, $P_{mn}(t)$, and $Q_{mn}(t)$ are respectively the current, the active and the reactive power flowing in the branch from the m-th bus to the n-th one, R_{mn} and X_{mn} are the resistance and reactance of the mn-th branch. $P_n^c(t)$ and $P_n^d(t)$ are the charging and discharging power of the storage at time t. $P_n^g(t)$ and $Q_n^g(t)$ are the active and reactive power provided by the upstream connections (slack bus of the network). The values of $P_n^g(t)$ and $Q_n^g(t)$ are zero except for the first node.

2.6. Network Constraints

The current magnitude quadratic term can be defined as the function of the corresponding active and reactive power quadratic terms (Equations (9)–(11)).

$$I_{mn}^2 \geq \frac{P_{mn}^2 + Q_{mn}^2}{V_m^2} \tag{9}$$

$$P_{mn}^2(t) + Q_{mn}^2(t) = S_l^2(t) \tag{10}$$

$$i_{mn}(t) \cdot v_m(t) = S_l^2(t) \tag{11}$$

Equation (9) is relaxed ultimately by relaxing the magnitude of currents within each branch and using a conic formation on the limitation of exchanged active power. For linearization purposes, the quadratic terms of voltage and current magnitude have been replaced with the linear ones as in (12).

$$I_{mn}^2 = i_{mn}; \quad V_m^2 = v_m \quad (12)$$

The new variables (i_{mn}, v_m) successfully formulate the SOCP problem according to the following constraint,

$$V_{min}^2 \leq v_m(t) \leq V_{max}^2 \quad (13)$$

The Equation (13) provides the voltage limits of each bus.
Bus 1 is modelled as a swing bus with fixed complex voltage $V_m(t)$.

2.7. Constraints for RES and Controllable Generator

Equations (14) and (15) impose the limits to the active and reactive power curtailment associated with RES and CHP generators. In Equation (14), $P_n^{min\ RESc/CHPc}$ represents the lower bound of the active power curtailment of RES and CHP generators. In this study, the lower bound value of curtailment has been chosen as 0, which means the generators curtail all of their capacity. The upper bound, $P_n^{maxRESc/CHPc}$, has been considered the capacity of the generators based on the expected values of each time step.

$$P_n^{min\ RESc/CHPc} \leq P_n^{RESc/CHPc}(t) \leq P_n^{maxRESc/CHPc} \quad (14)$$

$$Q_n^{min\ RESc/CHPc} \leq Q_n^{RESc/CHPc}(t) \leq Q_n^{maxRESc/CHPc} \quad (15)$$

Furthermore, the constraints about storages may be formulated as in (16)–(20).

$$SOC_n(t) = SOC_n(t-1) + \left(P_n^c(t)\cdot\eta_c - \frac{P_n^d(t)}{\eta_d}\right)\cdot \Delta t \quad (16)$$

$$0 \leq P_n^c(t) \leq \alpha_n^c \cdot P_n^{c,max}(t) \quad (17)$$

$$0 \leq P_n^d(t) \leq \alpha_n^d \cdot P_n^{d,max}(t) \quad (18)$$

$$SOC_{n,min} \leq SOC_n(t) \leq SOC_{n,max} \quad (19)$$

$$\alpha_n^c(t) + \alpha_n^d(t) \leq 1 \quad (20)$$

where $\alpha_n^c(t) \in [0 \text{ or } 1]$ and $\alpha_n^d(t) \in [0 \text{ or } 1]$.

The state of charge (SoC) of ESSs is calculated by considering the initial SoC and the charging and discharging efficiencies η_c and η_d (Equation (16)). To restrict the maximum charging and the depth of discharging and for avoiding the simultaneous charging and discharging, the binary variables α_n^c and α_n^d, of which only one can be different from zero, have been considered in Equations (17)–(20). Finally, Equation (21) is added to force the SoC to be equal at the beginning and the end of the considered time horizon T.

$$SOC_{n,0} = SOC_{n,T} \quad (21)$$

The multiplication of binary and integer variables during the estimation of the charging and discharging power of the storage unit generates a quadratic term. A decomposition technique has been used to linearize the relevant constraints by rewriting constraints in the form of (22) as in (23) and (24) to avoid the bilinear terms.

$$x <= y \cdot z \cdot c \quad (22)$$

$$x <= y \cdot z_{max} \cdot c_{max} \quad (23)$$

$$x <= z \cdot c \quad (24)$$

x and c are continuous, y binary, z integer. The continuous and integer variables are respectively variable in $[0, x_{max}]$, $[0, c_{max}]$ and $[0, z_{max}]$.

3. Uncertainty Management

Uncertainties are mostly involved in decision-making problems. The uncertainty from electric loads, wind, and solar power generation typically influence distribution planning in general and storage allocation in particular. Several factors determine the evolution of each uncertainty. For example, the consumers' activities, energy savings and electricity providers' rate policies influence the electric load; the radiation of the sun and the velocity of air impact on the power output of PV and wind [34].

A static robust optimization is used to consider the uncertainty in the optimal planning model. To define the uncertainty set, an interval uncertainty model has been adopted with the flexibility to regulate the robustness, called budget of uncertainty (Γ_i). Static robust optimization devising seeks for optimal solutions that optimize the objective function and encounter the problem requirements for every possible revealing of the uncertainty in constraint coefficients. Hence, the variables are independent of the uncertain parameters.

For a worst-case analysis, when considering the uncertainty, the following problem (25)–(27) is dealt with:

$$\min c \cdot x \qquad (25)$$

Subject to

$$\sum_{j=1}^{n} a_{ij} x_j + \max \sum_{j \in J_i} \widetilde{a}_{ij} \xi_{ij} x_j \leq b_j \qquad (26)$$

$$l \leq x \leq u \qquad (27)$$

In the above optimization problem, Equations (25) and (26) represent the objective function and inequality constraint, respectively. The uncertainty bound of the uncertain parameter x, that must assume values between lower l and upper u bounds, as formulated by the Equation (27). b_j is the value of the right-hand side of i-th constraint.

For the i-th constraint, the auxiliary problem can be formulated as follows:

$$\max \sum_{j \in J_i} \widetilde{a}_{ij} \xi_{ij} |x_j| \qquad (28)$$

Subject to

$$\sum_{j \in J_i} \xi_{ij} \leq \Gamma_i \qquad (29)$$

$$0 \leq \xi_{ij} \leq 1 \qquad (30)$$

To make the model tractable, that means to convert the inner maximization problem to a minimization problem, the dual of the above problem (28)–(30) needs to be formulated as follows:

$$\min z_i \Gamma_i + \sum_{j \in J_i} p_{ij} \qquad (31)$$

Subject to

$$z_i + p_{ij} \geq \widetilde{a}_{ij} y_i V_i, \; j \in J_i \qquad (32)$$

$$|x_j| \leq y_j \qquad (33)$$

$$z_i, p_{ij}, y_j \geq 0 \qquad (34)$$

where z_i, p_{ij} are the dual decision variables for constraints of the auxiliary problem.

Incorporating model (31)–(34) into the original problem (25)–(27), the robust linear counterpart is formulated as:

$$\min c \cdot x \qquad (35)$$

Subject to

$$\sum_{j=1}^{n} a_{ij} x_j + z_i \Gamma_i + \sum_{j \in J_i} p_{ij} \leq b_i \qquad (36)$$

$$l_j \leq x_j \leq u_j \qquad (37)$$

$$z_i + p_{ij} \geq \widetilde{a}_{ij} y_j, \; \forall i, j \in J_i \qquad (38)$$

$$-y_j \leq x_j \leq y_j \qquad (39)$$

$$z_i, p_{ij}, y_j \geq 0 \qquad (40)$$

4. Robust Counterpart

Assume that all the decision variables should be considered before the revealing of the uncertainty from solar power, wind generation, and electric loads. In the active power balance (7), uncertainties $P_n^{RES}(t)$ and $PD_n(t)$ are modeled as symmetric and bounded variables $\widetilde{P_n^{pv}}(t)$, $\widetilde{P_n^{w}}(t)$ and $\widetilde{PD_n}(t)$. It should be mentioned here that $P_n^{RES}(t)$ consists of the solar and wind generations. The uncertainty takes values as in the following Equations (41)–(43).

$$\widetilde{P_n^{pv}}(t) = P_n^{pv}(t) + \Delta \hat{P}_n^{pv}(t) \; P_{pv}^{\hat{lb}} \leq \Delta P_{pv}(t) \leq P_{pv}^{\hat{ub}} \qquad (41)$$

$$\widetilde{P_n^{wind}}(t) = P_n^{wind}(t) + \Delta \hat{P}_n^{wind}(t) \; P_{wind}^{\hat{lb}} \leq \Delta P_{wind}(t) \leq P_{wind}^{\hat{ub}} \qquad (42)$$

$$\widetilde{PD_n}(t) = PD_n(t) + \Delta \hat{PD}_n(t) \; P_D^{\hat{lb}} \leq \Delta P_D(t) \leq P_D^{\hat{ub}} \qquad (43)$$

In the robust model, the objective function (1) is identical to the deterministic model. The only constraint that is affected by uncertainty is the electric power balance equation. The electric power in the network should be met when the worst case of uncertainties occurs. For the power balance equation, the worst case would occur at the maximum increase of the electric loads and the maximum decrease in solar (PV) and wind power generation. Therefore, the robust formulation becomes as in (44)–(47).

$$\min C_{tot} = \min \left\{ \sum_{n=1}^{N} \left[C_n^{RESc} + C_n^{CHPc} + C_n^{PLS} + C_n^{CAPEX\_ESS} \right] \right\} \qquad (44)$$

Subject to

$$\begin{aligned} P_n^g(t) + P_n^{RES}(t) \quad & -P_n^{RESc}(t) + P_n^{CHP}(t) - P_n^{CHPc}(t) - PD_n(t) + P_n^{PLS}(t) \\ & -P_n^c(t) + P_n^d(t) + \max\{P_D^{ub}(t) * \xi_D^{ub}(t) + P_D^{lb}(t) * \xi_D^{lb}(t) \\ & -P_{pv}^{ub}(t) * \xi_{pv}^{ub}(t) - P_{pv}^{lb}(t) * \xi_{pv}^{lb}(t) - P_{wind}^{ub}(t) * \xi_{wind}^{ub}(t) \\ & -P_{wind}^{lb}(t) * \xi_{wind}^{lb}(t)\} - \sum_{m \in \theta_n} R_{mn} \cdot I_{mn}^2 = \sum_{m \in \theta_n} P_{mn} \end{aligned} \qquad (45)$$

$$\xi_D^{ub}(t) + \xi_D^{lb}(t) + \xi_{pv}^{ub}(t) + \xi_{pv}^{lb}(t) + \xi_{wind}^{ub}(t) + \xi_{wind}^{lb}(t) \leq \Gamma_1(t) \qquad (46)$$

$$\xi_D^{ub}(t), \xi_D^{lb}(t), \xi_{pv}^{ub}(t), \xi_{pv}^{lb}(t), \xi_{wind}^{ub}(t), \xi_{wind}^{lb}(t) \leq 1 \qquad (47)$$

where $\xi_D^{ub}(t), \xi_D^{lb}(t), \xi_{pv}^{ub}(t), \xi_{pv}^{lb}(t), \xi_{wind}^{ub}(t), \xi_{wind}^{lb}(t)$ are the scaled deviations from the random electric loads, solar, and wind power generation, respectively. $\Gamma_1(t)$ is the budget of the uncertainty of uncertain parameters at time t that lies between 0 to 1, where 0 being the deterministic case and 1 defined the most robust case.

To make tractable the above problem, the following subproblem in Equations (48)–(50) need to be formulated into the corresponding dual problem by introducing dual variables $\lambda_1(t)$, $\Pi_D^+(t)$, $\Pi_D^-(t)$, $\Pi_{pv}^+(t)$, $\Pi_{pv}^-(t)$, $\Pi_{wind}^+(t)$, $\Pi_{wind}^-(t)$ for constraints (49) and (50).

The subproblem can be formulated as in (48).

$$\max\{P_D^{ub}(t) * \xi_D^{ub}(t) + P_D^{lb}(t) * \xi_D^{lb}(t) - P_{pv}^{ub}(t) * \xi_{pv}^{ub}(t) - P_{pv}^{lb}(t) * \xi_{pv}^{lb}(t) \\ - P_{wind}^{ub}(t) * \xi_{wind}^{ub}(t) - P_{wind}^{lb}(t) * \xi_{wind}^{lb}(t)\} \tag{48}$$

Subject to

$$\xi_D^{ub}(t) + \xi_D^{lb}(t) + \xi_{pv}^{ub}(t) + \xi_{pv}^{lb}(t) + \xi_{wind}^{ub}(t) + \xi_{wind}^{lb}(t) \leq \Gamma_1(t) \tag{49}$$

$$\xi_D^{ub}(t), \xi_D^{lb}(t), \xi_{pv}^{ub}(t), \xi_{pv}^{lb}(t), \xi_{wind}^{ub}(t), \xi_{wind}^{lb}(t) \leq 1 \tag{50}$$

The robust counterpart after applying the duality theory is formulated as in (51)–(55).

$$\min \lambda_1(t)\Gamma_1(t) + \Pi_D^+(t) + \Pi_D^-(t) + \Pi_{pv}^+(t) + \Pi_{pv}^-(t) + \Pi_{wind}^+(t) + \Pi_{wind}^-(t) \tag{51}$$

Subject to

$$\lambda_1(t) + \Pi_D^+(t) \geq \hat{P}_D^{ub}(t), \ \lambda_1(t) + \Pi_D^-(t) \geq \hat{P}_D^{lb}(t) \tag{52}$$

$$\lambda_1(t) + \Pi_{pv}^+(t) \geq -\hat{P}_{pv}^{ub}(t), \ \lambda_1(t) + \Pi_{pv}^-(t) \geq -\hat{P}_{pv}^{lb}(t) \tag{53}$$

$$\lambda_1(t) + \Pi_{wind}^+(t) \geq -\hat{P}_{wind}^{ub}(t), \ \lambda_1(t) + \Pi_{wind}^-(t) \geq -\hat{P}_{wind}^{lb}(t) \tag{54}$$

$$\lambda_1(t), \Pi_D^{\pm}(t), \Pi_{pv}^{\pm}(t), \Pi_{wind}^{\pm}(t) \geq 0 \tag{55}$$

Finally, the tractable robust model can be formulated as the following (56) and (57).

$$\min C_{tot} = \min\left\{\sum_{n=1}^{N}[C_n^{RESc} + C_n^{CHPc} + C_n^{DR} + C_n^{CAPEX_{ESS}}]\right\} \tag{56}$$

Subject to

$$P_n^g(t) + P_n^{RES}(t) - P_n^{RESc}(t) + P_n^{CHP}(t) - P_n^{CHPc}(t) - PD_n(t) + P_n^{PLS}(t) - \\ P_n^c(t) + P_n^d(t) + \lambda_1(t)\Gamma_1(t) + \Pi_D^+(t) + \Pi_D^-(t) + \Pi_{pv}^+(t) + \Pi_{pv}^-(t) + \Pi_{wind}^+(t) + \\ \Pi_{wind}^-(t) - \sum_{m \in \theta_n} R_{mn} \cdot I_{mn}^2 = \sum_{m \in \theta_n} P_{mn}(t) \tag{57}$$

Moreover, the constraints (8)–(21) and (52)–(55) form the tractable problem.

The new model does not contain any uncertainty and is formulated as a mixed-integer second-order conic programming (MISOCP) problem that can be solved efficiently using CPLEX that uses a branch and cut algorithm to find the integer feasible solution.

5. Case Study

The procedure was applied to a test distribution network derived from the ATLANTIDE project [32]. The MV network, shown in Figure 1, representative of the industrial ambit, was constituted by 100 nodes, subdivided in seven feeders supplied by a primary substation equipped with a 25 MVA high voltage/medium voltage (HV/MV) transformer. The total demand was about 30 MVA (372 GWh/year) and the total installed DG capacity was 34 MW (27.2 GWh/year), as a mix of wind, PV and biomass CHP generators.

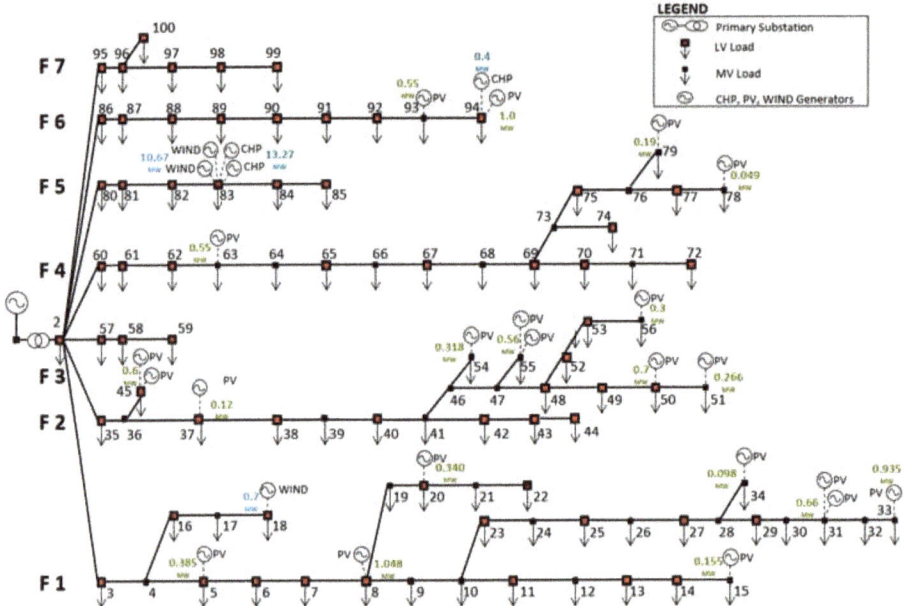

Figure 1. Test network (representative network derived from ATLANTIDE project [32]).

The mathematical formulation of the RO for an AC-OPF based energy storage planning tool was programmed in General Algebraic Modeling System (GAMS) (GAMS Software GmbH, Frechen, Germany) and solved using CPLEX 25.1.1 on a 2.30 GHz personal computer with 4 GB RAM. In this experimental study, the worst case was considered when the load was high ($\xi_{D,t} = 1$) and wind and PV generation was low ($\xi_{pv,t}, \xi_{w,t} = -1$).

For the sake of a comprehensive view, in the following, the results obtained by the application of the described optimization to the network of Figure 1 in 12 typical days, differentiated between working days, Saturdays, and holidays (Sundays included), and between seasons, have been reported. The time horizon of 24 h of each typical day has been considered with a time step of 1 h. Three scenarios have been considered: the certain one (solved by the deterministic OPF) and two uncertain scenarios with different values of risk ($\Gamma = 0.5$ and $\Gamma = 1$), both solved with RO. Furthermore, for highlighting the advantages provided by the storage systems the case of deterministic optimization (certain) without storage has been added to the previously described cases.

All the buses of the test network were assumed candidates for storage placement. The available ESS were considered of 1.0 MW/2 h storage capacity. The efficiencies for charging and discharging were considered 90% each, which gives an overall roundtrip efficiency of 0.81. The initial state of charge (SoC) has been considered 25% of its capacity.

In these typical days, some under-voltage conditions occur in the most distant nodes from the HV/MV transformer and, thus, for solving these issues, it is necessary to resort to the load peak shaving. Furthermore, some lines suffer for overloading depending on the non-coincidence of load demand and DG production. ESSs prove to be useful for reducing the curtailment of the demand and production as detailed in the next subsections.

5.1. Generation and Load Profiles

The generation and load profiles were simulated according to the ATLANTIDE load and generation daily curves, that provide for different kinds of customers (i.e., industrial, residential, commercial, and agricultural) and for several technologies of DG (i.e., wind turbine, PV, and CHP biomass-based) the hourly consumption/production for each typical day. An amount of 22 PV systems was assigned to 20 nodes. The size of these systems is between 49–1048 kW. Node 8 had the biggest PV system, whereas the lowest one was connected to node 78. Node 83 comprised two wind generators and two CHP plants. Figure 1 depicts the nominal power of the PV, wind, and CHP of each node.

The load profiles indicated a peak load of 18.69 MW during the spring working day and 18.14 MW during the summer working day with an average load of 13.79 MW and 13.38 MW, respectively. As an example, the demand and production profiles and their balance at the HV/MV interface, during the spring working day, are shown in Figure 2.

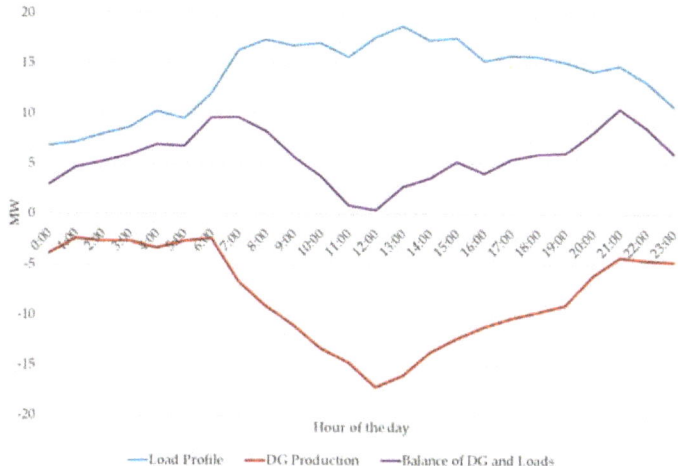

Figure 2. Load-production profiles of the whole network for the Spring working typical day.

5.2. Storage Placement

The optimization results of storage position for each typical day for the three considered cases have been enumerated in Table 1. The ESS optimal positions may change from one typical day to another even for the same case, but the results can be summarized by considering a given solution valid for all the twelve typical days. In the following, it has been assumed that the placement in one bus or a close one on two different typical days can be considered the same placement. For instance, the bus 83 and the bus 84 in the deterministic case, that are the solutions for the TD8, and the TD5 respectively, can be considered as a unique optimal position around the bus 83.

Table 1. Storage placement for each typical day for the three considered cases.

Typical Days	Deterministic Case	Intermediate Case	Robust Case
TD1 (Winter working day)	10, 32	10, 32, 77	10, 32, 48, 77
TD2 (Winter Saturday)	-	-	-
TD3 (Winter holiday)	-	-	-
TD4 (Spring working day)	10	10, 34	10, 34
TD5 (Spring Saturday)	84	83, 84	83, 84
TD6 (Spring holiday)	84	84	84
TD7 (Summer working day)	-	-	-
TD8 (Summer Saturday)	83	12, 83	10, 32, 83
TD9 (Summer holiday)	-	-	-
TD10 (Autumn working day)	12	12, 27	12, 27, 69
TD11 (Autumn Saturday)	84	84	69, 84
TD12 (Autumn holiday)	83, 85	83, 85	83, 85
Total number of ESS	4	5	6

On the contrary, if two busses, even close, appeared in the solution of the same day they were both considered necessary and two ESSs had to be placed on that nodes (e.g., the busses 83 and 85 in the results of all the cases for the TD12 or the busses 83 and 84 in the results of the intermediate and robust cases for the TD5). By applying these rules, the total number of ESSs that had to be placed in the three cases are reported in the last row of Table 1. It is worthy of mentioning that the results were substantially incremental: the intermediate case included the location of the deterministic case, and the robust case (no risk) included, in turn, the intermediate one.

To analyze the impact of renewables and load uncertainty on the investment of the energy storage in the distribution network, one of the worst-cases of RES (PV, wind or biomass based) and combination of loads were considered. The worst-case scenario considered in this work was when the loads had upper bound values, and the renewables had lower bound values. Three cases were considered by varying the loads and renewables uncertainty bounds. In the first case, the budget of uncertainty was zero ($\Gamma = 0$), i.e., the profiles of load and renewable generations were assumed following the forecasted values. In the second case, the value of budget of uncertainty for both load and renewables considered 0.5 ($\Gamma = 0.5$) that is between the zero (deterministic) and 1 (robust or worst case). In the third case ($\Gamma = 1$), the considered worst-case scenario was evaluated. In this case, the uncertainty sets of loads and renewables were considered broader to consider the possible extreme coordinates of the uncertainty set.

The following figures compare the results of the studied cases (i.e., no control, deterministic OPF no storage, deterministic OPF with storage, intermediate and robust). These results are related to the most critical typical day, the winter working day (TD1). For the sake of clarity, the figures refer only to the feeder F1 that is the longest feeder of the test network depicted in Figure 1 (i.e., the last bus is about 14.2 km far from the primary substation). Figure 3 shows the voltage profiles occurring at 9:00 am of the winter working typical day, because, among other time intervals, this one was proved that experiments the greatest load curtailment; Figure 4 shows the load curtailed during this typical day, in Figure 5 the balances of DG production and curtailed demand, and, finally, Figure 6 the ESS charging/discharging optimal profiles of one of the ESS optimal positioned in the feeder F1 (bus 10 of Figure 1). As it is evident by the results, all the optimizations allow to solve the undervoltage conditions occurring in the long feeder F1 (Figure 3); the more conservative the optimization (i.e., by moving from certain to uncertain, intermediate and robust, optimization) the smaller the demand curtailed (Figure 4); in the feeder F1 no generation curtailment results from the optimizations, thus the balance of production and demand (curtailed) is closer to the original one (no control) in the robust case (Figure 5). It is worth noticing that the voltage value at the sending end (the MV busbar of the primary substation) was lower in the no control case than the other cases because the implemented model of the HV/MV transformer is very simple and strongly suffers for the high demand, not curtailed in the control case. In future works, the transformer model will be improved. These results, together with the ESS operation, are discussed more in detail in the next subsections for each optimization case.

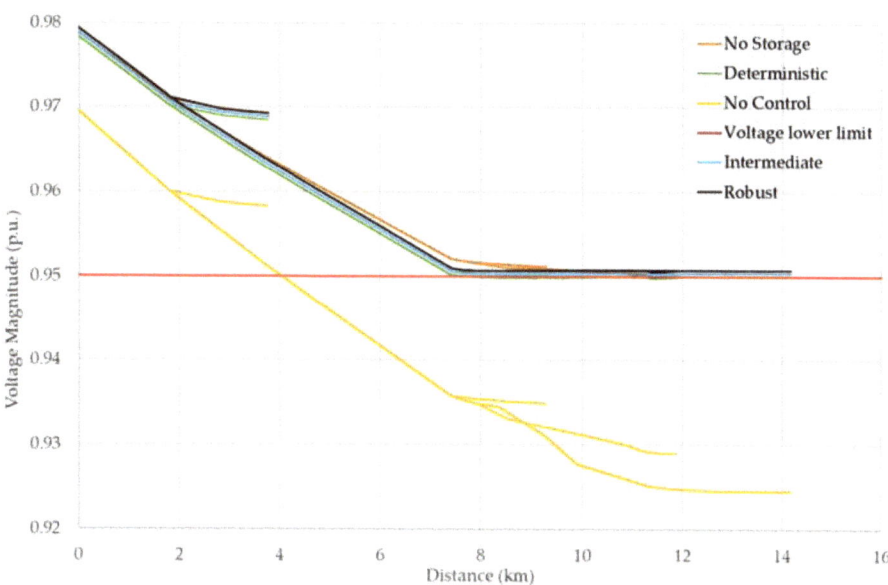

Figure 3. Voltage profiles of feeder F1 for no control, no storage (certain deterministic optimal power flow (OPF) without storage) and deterministic (certain deterministic OPF with storage), intermediate and robust cases at 9:00 am of the winter working day.

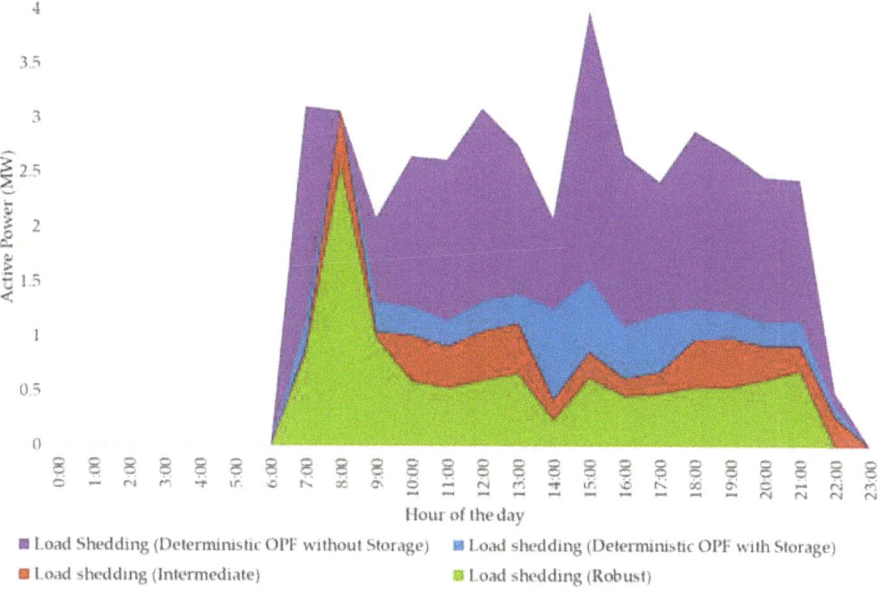

Figure 4. Load curtailments experimented by the feeder F1 for the no storage (certain deterministic OPF without storage), deterministic (certain deterministic OPF with storage), intermediate and robust cases on the winter working day.

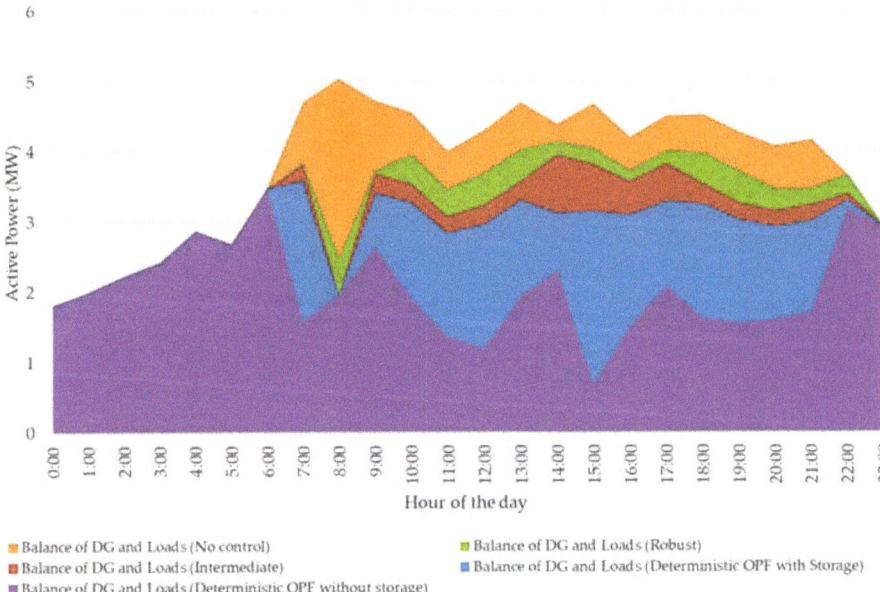

■ Balance of DG and Loads (No control)   ■ Balance of DG and Loads (Robust)
■ Balance of DG and Loads (Intermediate) ■ Balance of DG and Loads (Deterministic OPF with Storage)
■ Balance of DG and Loads (Deterministic OPF without storage)

Figure 5. Balance of distributed generation (DG) production and curtailed demand of the feeder F1 for the no control, no storage (certain deterministic OPF without storage), deterministic (certain deterministic OPF with storage), intermediate and robust cases on the winter working day.

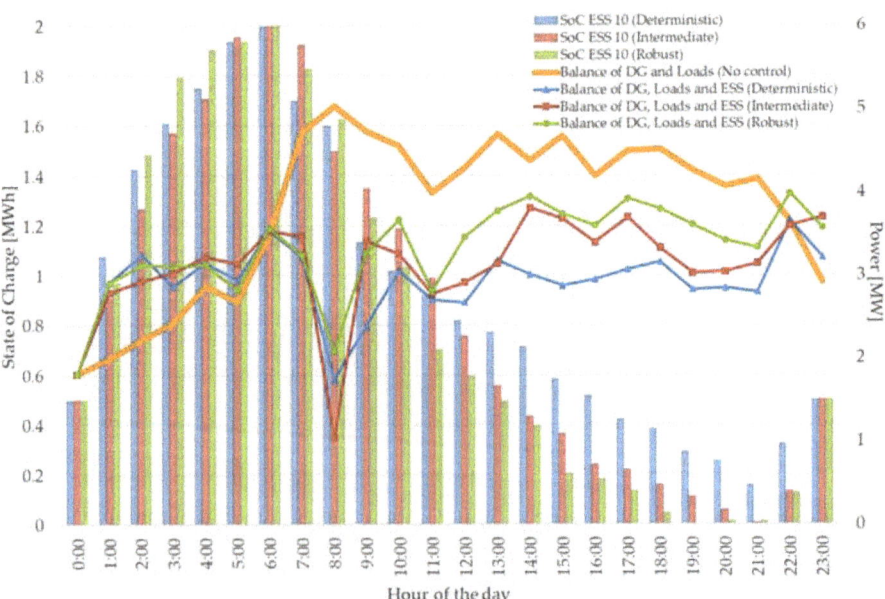

Figure 6. Charging/discharging profiles of the energy storage system (ESS) optimally positioned in bus 10 of the feeder F1 for the deterministic, intermediate and robust cases on the winter working day and balances of powers (DG, loads, and ESS) in the same cases. The no control case has been added for comparison.

5.3. Deterministic Case (with and without Storage)

Due to the absence of uncertainty, in this case, the load and renewables profiles will remain the same as the predicted values. It was witnessed that during the deterministic case with storage, at least four storages need to cover their requirements. The ESS optimal positions can change from one typical day to another, but, by summarizing the results in the 12 typical days (Table 1), they were located two in the two lateral branches that start from the node 10 (feeder F1, positions are 10 or 12 and 32). The third and fourth ESS had to be located around the node 83 (feeder F5, positions are two among 83, 84 and 85). It is essential to observe that node 83 is the node that had the highest number of renewables and CHP connected; thus, it is noticeable to consider that as a privileged position for storages.

The highest load curtailment was experienced during the typical day of winter working day (TD1) with the amount of 51.77 MWh/day for the case without storage and 30.46 MWh/day for the case with storage.

By focusing on the feeder F1, as it is evident from Figure 3, the nodes of this feeder had under-voltage issues in the no control case, and any optimization forces to resort load shedding (Figure 4). By comparing these two certain cases, is it worth noticing that if the ESSs are not available for the optimization (deterministic OPF no storage) much more demand had to be curtailed (i.e., 41.62 MWh/day of the no storage case vs. 20.97 MWh/day in the case with storage).

The daily operation of the ESS was optimized as well as the optimal position. For instance, during the winter working day, the daily operation profile of the ESS located around bus 10 is shown in Figure 6, together with the balances of demand and production curves, with and without ESS. At the beginning of the day, the ESS started to charge, keeping the final balance of demand, DG production (minimal in the first hours of the day), and charging power for ESS so low to do not negatively impact the network operation. At around 7:00 am, when the morning peak starts, the storage discharges for reducing the power demand and keeping the voltage profile within the limit (Figure 3).

5.4. Intermediate Case

In this case, a narrow uncertainty bound is considered. The budget of uncertainty for the uncertain parameters has been considered as $\Gamma = 0.5$. The optimization algorithm will look for a solution inside the specified uncertainty bound. From Table 1, by considering the simulation results of the twelve typical days, and assuming the most conservative hypotheses (i.e., the final result is the union of the results obtained for each typical day), the intermediate case suggests at least five storage systems to be installed: two in the feeder F1 and two in the feeder F5, as in the deterministic case, plus one ESS in the feeder F4. The positions of the two storages in the feeder F1 and the two in the feeder F5 are more or less the same of the deterministic case (F1 possible locations are the busses 10 or 12 for one lateral and the busses 27, 32, or 34 for the other lateral, and two positions among the bus 83, 84 or 85 for the feeder F5). In the feeder F4, the added storage system has to be installed around the node 77.

The load shedding, in this case, is more reduced and for the critical TD1 is equal to 15.89 MWh/day (about 25% less than the deterministic case), as shown in Figure 4.

Furthermore, it can be observed in Figure 6 that the ESS located around bus 10 in the feeder F1 has a similar trend of the same ESS in the deterministic case: it charges and discharges mostly in the same hours for solving local contingencies. In particular, it charges when the load demand is low (at the first hours of the day), and finally, at the end of the day for recovering their initial SoC; on the contrary, it discharges in correspondence of the peaks of demand (7:00–21:00).

5.5. Robust Case

The third case can be considered the worst-case analysis. In this case, the budget of uncertainty for uncertain parameters is equal to 1 ($\Gamma = 1$). This budget of uncertainty allows the algorithm to consider the extreme points of the uncertainty set. Compared to the previous cases, the robust case provides six storage systems to be installed in the network. The locations of storage for feeder F1, F4 and F5 are like the deterministic and intermediate cases. However, the robust case suggests one more ESS in the feeder F2 (bus 48). In Figure 6, the operation profile of the ESS located around bus 10, resulting from the optimization for the feeder F1 is shown with the demand and production daily curves. The behavior of the ESS is as the one in the other cases: the contribution to reducing the peaks at the cost of a slight increase in demand when they charge. This increase does not alter the network operation and does not produce any technical constraint violation, but it allows a further reduction of load shedding (Figure 4). For the feeder 1 in the critical TD1, the demand is curtailed of 11.10 MWh/day (−30% than the intermediate case).

5.6. Economic Analysis

In order to analyze the economic feasibility of the investments in storage systems, the comparison between all the cases mentioned above, included the no storage one, has been considered. Table 2 summarizes the yearly operational costs (operational expenditures—OPEX) for the four considered cases, the amount of load shedding and generation curtailment used for solving the contingencies, the CAPEX for the ESS installation referred to one year only (among the ten years of the ESS life duration), and in the last column, the total yearly cost is calculated as the summation of CAPEX and the OPEX. In the no storage case, the yearly operational cost, of about 1480 k€, consists of penalty cost for load shedding that accounts for 762.90 k€/year, and penalty cost of CHP curtailment worth 717.46 k€/year. The peak shaving drastically decreases by using the ESS even in the deterministic case (the quantity is about halved), and then it is significantly further reduced in the uncertain scenarios. The same behavior can be observed for the generation curtailment of CHP. In the uncertain cases, compared with the base case without storages, the resort to load shedding is much reduced (−44.8% in the deterministic case becomes −60.5% in the intermediate case and −73.9% in the robust one) as well as the generation curtailment (−22.8%, −43.9% and −66.3% in the deterministic, intermediate and robust case respectively). The quantities related to the generation curtailment in Table 2 for these four cases are referred only to the curtailment of CHPs.

Table 2. Daily operational cost of the test network and ESS CAPEX.

Cases	OPEX [k€/Year]	Load Shedding [MWh/Year]	Generation Curtailment [MWh/Year]	CAPEX [k€/Year]	Total Cost CAPEX+OPEX [k€/Year]
No storage	1480.36	6602.12	7502.38	0	1480.36
Deterministic ($\Gamma = 0$)	974.47	3644.95	5791.67	400	1374.47
Intermediate ($\Gamma = 0.5$)	704.64	2605.34	4208.2	500	1204.64
Robust ($\Gamma = 1$)	437.80	1721.07	2525.22	600	1037.80

Consequently, a substantial reduction of the annual operational costs can be observed with the ESS inclusion in the deterministic case and much more in the uncertain cases (−34.2%, −52.4%, and −70.4%, in the deterministic, intermediate and robust case respectively). It is worth noticing that, in the deterministic and uncertain cases, apart from the operational costs, an additional cost factor has to be considered: the CAPEX for the ESS installation, split in ten years (the CAPEX of one 1.0 MW/2 h ESS is assumed the same for each year). This negatively impacts on the final cost, much stronger with the increment of budget or uncertainty, due to the growth of the investment costs for the increasing number of storages. However, the reduction of OPEX not only covers such increase but, the final costs of all the cases that use the storage systems for relieving the contingencies are smaller than the case without them (no storage case). In particular, the percentage of total cost reduction is smaller than the one calculated by considering the OPEX only (i.e., −7.2%, −18.6% and −29.9% for deterministic, intermediate and robust cases, respectively), but the results prove the effectiveness of the optimization. In fact, these results demonstrate that not only the ESS helps to reduce the operational cost, for relieving even the worst-case and reducing even more the resort to load shedding and to the generation curtailment, but also that, with the assumed hypotheses, the ESS CAPEX can be amortized during the ten years of their life duration.

6. Conclusions

This paper establishes the use of a SOCP convex relaxation of the power flow equations for optimal placement of energy systems in an MV distribution network. The algorithm proposed in this paper can be used to analyze the economic viability in comparison to investment and operational costs. The application of robust optimization and having the flexibility to modulate the budget of uncertainty helps to find a balance among the factors of economic efficiency and conservatism. The use of this kind of flexibility also assisted in considering additional scenarios other than the worst-case scenarios that most robust optimization problems account for. By considering the worst-case scenario only, such problems do not provide an optimal solution. Rather, they offer only conservative solutions that could be impractical. However, the analytical reformulation technique helped to find the robust equivalent of the original problem that was solved with less computational encumbrance using CPLEX solver.

As planning includes a limited financial budget and resources, this study affords a comprehensive approach, which is a consideration of different situations (budget of uncertainty). Furthermore, the use of this innovative algorithm leads to understanding of the benefits of grid-connected storage devices in distribution systems and the consideration of uncertainties into the planning phase.

In future works, the implementation of the HV/MV transformer model will be improved. Moreover, the reactive power provision from storage will be considered in the future model. A term that takes into account the depreciation of the ESSs due to their use will be included.

Author Contributions: Conceptualization, N.C., G.P., and F.P.; methodology, N.C.; software, N.C.; validation, G.P. and N.C.; formal analysis, F.P., G.P., and N.C.; investigation, G.P. and N.C.; resources, N.C.; data curation, N.C. and G.P.; writing—original draft preparation, N.C. and G.P.; writing—review and editing, N.C. and G.P.; visualization, N.C.; supervision, G.P.; project administration, F.P.; funding acquisition, F.P. All authors have read and agreed to the published version of the manuscript.

Funding: Nayeem Chowdhury has been funded from the European Union's Horizon 2020 research and innovation programme under Grant Agreement No 676042. The contribution of G. Pisano to this paper has been conducted within the R&D project "Cagliari2020" partially funded by the Italian University and Research Ministry (grant# MIUR_ PON04a2 _00381).

Conflicts of Interest: The authors declare no conflict of interest.

Nomenclature

c_E, c_P	Specific costs of energy storage in terms of energy and rated power, respectively
c_{EN}	Energy price in the wholesale market
$C_n^{CAPEX\_ESS}$	Storage investment cost (CAPEX, capital expenditures) at node n
C_n^{CHPc}	Cost of curtailing combined heat and power (CHP) power generation at node n
C_n^{PLS}	Cost of peak load shaving at node n
C_n^{REsc}	Cost of renewable energy curtailment at node n
F	Fuel cost for biomass CHP plant
$I_{mn}(t)$	Current flows in the branch from m-th to the n-th bus at the time interval t
K_S	Capital recovery factor
$PD_n(t)$, $QD_n(t)$	Active and reactive power demand of the loads at node n during the time interval t, respectively
$P_{mn}(t)$, $Q_{mn}(t)$	Active and reactive power flows in the branch from m-th to the n-th bus at the time interval t, respectively
$P_n^c(t)$, $P_n^d(t)$	Charging and discharging power of storage at node n during the time interval t, respectively

Symbol	Description
$P_n^{c,max}(t), P_n^{d,max}(t)$	Maximum and minimum limits of charging and discharging power of storage at node n during the time interval t, respectively
$P_n^{CHP}(t), Q_n^{CHP}(t)$	Expected active and reactive power production of CHP at node n during the time interval t, respectively
$P_n^g(t), Q_n^g(t)$	Active and reactive power provided by the upstream connections at node n during the time interval t, respectively
$P_n^{PLS}(t)$	Power related to the peak load shaving at node n during the time interval t
$P_n^{RESc}(t)$	Amount of renewable power curtailment at node n during the time interval t
$P_n^{RES}(t), Q_n^{RES}(t)$	Expected active and reactive power production of renewables at node n during the time interval t, respectively
$P_n^{maxRESc/CHPc}$	Upper bound of active power curtailment of renewables and CHP at node n
$P_n^{min\ RESc/CHPc}$	Lower bound of active power curtailment of renewables and CHP at node n
$\widetilde{P_n^{pv}}(t), \widetilde{P_n^{wind}}(t), \widetilde{PD_n}(t)$	Bounded variables of PV, wind and power demand of loads at node n during time interval t, respectively
$\widetilde{\Delta P_n^{pv}}(t), \widetilde{\Delta P_n^{wind}}(t), \widetilde{\Delta PD_n}(t)$	Deviation from expected power value of PV, wind and power demand of loads at node n during time interval t, respectively
$Q_n^{min\ RESc/CHPc}$	Lower bound of reactive power curtailment of renewables and CHP at node n
$Q_n^{max\ RESc/CHPc}$	Upper bound of reactive power curtailment of renewables and CHP at node n
R_{mn}	Resistance of the mn-th branch
$S_l(t)$	Thermal capacity of the line at time interval t
$SOC_n(t)$	State of the charge of storage unit at node n during the time interval t
SC_n	The storage investment cost
V_{max}, V_{min}	Maximum and minimum voltage limits, respectively
X_{mn}	Reactance of the mn-th branch
$\alpha_n^c(t), \alpha_n^d(t)$	Binary variables for charging and discharging of storage at node n during the time interval t, respectively
$\Pi_D^+(t)\Pi_D^-(t), \Pi_{pv}^+(t), \Pi_{pv}^-(t)$	Dual variables of load and PV at the time interval t
$\Pi_{wind}^+(t), \Pi_{wind}^-(t)$	Dual variables of wind at the time interval t
η_c, η_d	Charging and discharging efficiency of storage, respectively
$\xi_D^{ub}(t), \xi_D^{lb}(t), \xi_{pv}^{ub}(t), \xi_{pv}^{lb}(t)$	Scaled deviations from the random electric loads and PV at the time interval t
$\xi_{wind}^{ub}(t), \xi_{wind}^{lb}(t)$	Scaled deviations from the random wind power generation at the time interval t
Γ_i	The budget of uncertainty of the uncertain parameter i

References

1. International Renewable Energy Agency. *REmap: Roadmap for A Renewable Energy Future, 2016 ed.*; IRENA: Abu Dhabi, UAE, 2016.
2. Dunn, B.; Kamath, H.; Tarascon, J.M. Electrical energy storage for the grid: A battery of choices. *Science* **2011**, *334*, 928–935. [CrossRef] [PubMed]

3. Dvorkin, Y.; Lubin, M.; Backhaus, S.; Chertkov, M. Uncertainty sets for wind power generation. *IEEE Trans. Power Syst.* **2015**, *31*, 3326–3327.
4. Pandžić, H.; Wang, Y.; Qiu, T.; Dvorkin, Y.; Kirschen, D.S. Near-optimal method for siting and sizing of distributed storage in a transmission network. *IEEE Trans. Power Syst.* **2014**, *30*, 2288–2300.
5. Gayme, D.; Topcu, U. Optimal power flow with large-scale storage integration. *IEEE Trans. Power Syst.* **2012**, *28*, 709–717. [CrossRef]
6. Makarov, Y.V.; Du, P.; Kintner-Meyer, M.C.; Jin, C.; Illian, H.F. Sizing energy storage to accommodate high penetration of variable energy resources. *IEEE Trans. Sustain. Energy* **2011**, *3*, 34–40. [CrossRef]
7. Grainger, B.M.; Reed, G.F.; Sparacino, A.R.; Lewis, P.T. Power electronics for grid-scale energy storage. *Proc. IEEE* **2014**, *102*, 1000–1013. [CrossRef]
8. Vargas, L.S.; Bustos-Turu, G.; Larraín, F. Wind power curtailment and energy storage in transmission congestion management considering power plants ramp rates. *IEEE Trans. Power Syst.* **2014**, *30*, 2498–2506.
9. Chowdhury, N.; Pilo, F.; Pisano, G.; Troncia, M. Optimal Location of Energy Storage Systems with Robust Optimization. In Proceedings of the CIRED 2019-25th International Conference and Exhibition on Electricity Distribution, Madrid, Spain, 3–6 June 2019.
10. Bucciarelli, M.; Paoletti, S.; Vicino, A. Optimal sizing of energy storage systems under uncertain demand and generation. *Appl. Energy* **2018**, *225*, 611–621. [CrossRef]
11. Zhao, H.; Wu, Q.; Huang, S.; Guo, Q.; Sun, H.; Xue, Y. Optimal siting and sizing of Energy Storage System for power systems with large-scale wind power integration. In Proceedings of the 2015 IEEE Eindhoven PowerTech, Eindhoven, The Netherlands, 29 June–2 July 2015.
12. Wogrin, S.; Gayme, D.F. Optimizing storage siting, sizing, and technology portfolios in transmission-constrained networks. *IEEE Trans. Power Syst.* **2014**, *30*, 3304–3313.
13. Peker, M.; Kocaman, A.S.; Kara, B.Y. Benefits of transmission switching and energy storage in power systems with high renewable energy penetration. *Appl. Energy* **2018**, *228*, 1182–1197. [CrossRef]
14. Motalleb, M.; Reihani, E.; Ghorbani, R. Optimal placement and sizing of the storage supporting transmission and distribution networks. *Renew. Energy* **2016**, *94*, 651–659. [CrossRef]
15. Celli, G.; Pilo, F.; Pisano, G.; Soma, G.G. Distribution energy storage investment prioritization with a real coded multi-objective Genetic Algorithm. *Electr. Power Syst. Res.* **2018**, *163*, 154–163. [CrossRef]
16. Celli, G.; Chowdhury, N.; Pilo, F.; Soma, G.G.; Troncia, M.; Gianinoni, I.M. Multi-Criteria Analysis for decision making applied to active distribution network planning. *Electr. Power Syst. Res.* **2018**, *164*, 103–111. [CrossRef]
17. Paudyal, S.; Canizares, C.A.; Bhattacharya, K. Three-phase distribution OPF in smart grids: Optimality versus computational burden. In Proceedings of the 2011 2nd IEEE PES International Conference and Exhibition on Innovative Smart Grid Technologies, Manchester, UK, 5–7 December 2011.
18. Bose, S.; Low, S.H.; Teeraratkul, T.; Hassibi, B. Equivalent relaxations of optimal power flow. *IEEE Trans. Autom. Control* **2015**, *60*, 729–742. [CrossRef]
19. Gan, L.; Li, N.; Topcu, U.; Low, S.H. Exact convex relaxation of optimal power flow in radial networks. *IEEE Trans. Autom. Control* **2015**, *60*, 72–87. [CrossRef]
20. Papaefthymiou, S.V.; Papathanassiou, S.A. Optimum sizing of wind-pumped-storage hybrid power stations in island systems. *Renew. Energy* **2014**, *64*, 187–196. [CrossRef]
21. Fossati, J.P.; Galarza, A.; Martín-Villate, A.; Fontan, L. A method for optimal sizing energy storage systems for microgrids. *Renew. Energy* **2015**, *77*, 539–549. [CrossRef]
22. Balducci, P.J.; Alam, M.J.E.; Hardy, T.D.; Wu, D. Assigning value to energy storage systems at multiple points in an electrical grid. *Energy Environ. Sci.* **2018**, *11*, 1926–1944. [CrossRef]
23. Fernández-Blanco, R.; Dvorkin, Y.; Xu, B.; Wang, Y.; Kirschen, D.S. Optimal energy storage siting and sizing: A WECC case study. *IEEE Trans. Sustain. Energy* **2016**, *8*, 733–743. [CrossRef]
24. Jayasekara, N.; Masoum, M.A.S.; Wolfs, P.J. Optimal operation of distributed energy storage systems to improve distribution network load and generation hosting capability. *IEEE Trans. Sustain. Energy* **2016**, *7*, 250–261. [CrossRef]
25. Babacan, O.; Torre, W.; Kleissl, J. Siting and sizing of distributed energy storage to mitigate voltage impact by solar PV in distribution systems. *Sol. Energy* **2017**, *146*, 199–208. [CrossRef]
26. Hassan, A.; Dvorkin, Y. Energy storage siting and sizing in coordinated distribution and transmission systems. *IEEE Trans. Sustain. Energy.* **2018**, *9*, 1692–1701. [CrossRef]

27. Wang, H.; Lv, Q.; Yang, G.; Geng, H. Siting and sizing method of energy storage system of microgrid based on power flow sensitivity analysis. *J. Eng.* **2017**, *2017*, 1974–1978. [CrossRef]
28. Erdinc, O.; Tascikaraoglu, A.; Paterakis, N.G.; Dursun, I.; Sinim, M.C.; Catalao, J.P. Comprehensive optimization model for sizing and siting of DG units, EV charging stations and energy storage systems. *IEEE Trans. Smart Grid.* **2017**, *9*, 3871–3882. [CrossRef]
29. Castillo, A.; Gayme, D.F. Grid-scale energy storage applications in renewable energy integration: A survey. *Energy Convers. Manag.* **2014**, *87*, 885–894. [CrossRef]
30. Blanco, H.; Faaij, A. A review at the role of storage in energy systems with a focus on Power to Gas and long-term storage. *Renew. Sustain. Energy Rev.* **2018**, *81*, 1049–1086. [CrossRef]
31. Lorente, J.L.; Liu, X.A.; Best, R.; Morrow, D.J. Energy storage allocation in power networks–A state-of-the-art review. In Proceedings of the 2018 53rd International Universities Power Engineering Conference (UPEC), Glasgow, UK, 4–7 September 2018.
32. Bracale, A.; Caldon, R.; Celli, G.; Coppo, M.; Dal Canto, D.; Langella, R.; Petretto, G.; Pilo, F.; Pisano, G.; Proto, D. Analysis of the Italian distribution system evolution through reference networks. In Proceedings of the 2012 3rd IEEE PES Innovative Smart Grid Technologies Europe, ISGT Europe, Berlin, Germany, 14–17 October 2012.
33. Italian National Energy Market Operator (Gestore mercati energetici). Available online: https://www.mercatoelettrico.org/En/default.aspx (accessed on 20 January 2020).
34. Birge, J.R.; Louveaux, F. *Introduction to Stochastic Programming*; Springer Science & Business Media: Berlin, Germany, 2011.

© 2020 by the authors. Licensee MDPI, Basel, Switzerland. This article is an open access article distributed under the terms and conditions of the Creative Commons Attribution (CC BY) license (http://creativecommons.org/licenses/by/4.0/).

Article

Coordinated Operation of Energy Storage Systems for Distributed Harmonic Compensation in Microgrids

Abbas Marini [1], Luigi Piegari [2], S-Saeedallah Mortazavi [3] and Mohammad-S Ghazizadeh [1,*]

1. Department of Electrical Engineering, Abbaspour College, Shahid Beheshti University, Tehran 1983969411, Iran; a_marini@sbu.ac.ir
2. Department and Electronics, Information & Bioengineering, Politecnico di Milano, 20133 Milan, Italy; luigi.piegari@polimi.it
3. Department of Engineering, Shahid Chamran University of Ahvaz; Ahvaz 6135783151, Iran; mortazavi_s@scu.ac.ir
* Correspondence: ms_ghazizadeh@sbu.ac.ir; Tel.: +98-21-7393-2531

Received: 17 December 2019; Accepted: 7 February 2020; Published: 10 February 2020

Abstract: Energy storage systems (ESSs) bring various opportunities for a more reliable and flexible operation of microgrids (MGs). Among them, energy arbitrage and ancillary services are the most investigated application of ESSs. Furthermore, it has been shown that some other services could also be provided by ESSs such as power quality (PQ) improvements. This issue could be more challenging in MGs with widespread nonlinear loads injecting harmonic currents to the MG. In this paper, the feasibility of ESSs to act as coordinated active harmonic filters (AHF) for distributed compensation was investigated. An optimization model was proposed for the coordination of the harmonic compensation activities of ESSs. The model takes into account the various technical and systematic constraints to economically determine the required reference currents of various AHFs. Simulation cases showed the performance of the proposed model for enhancing the harmonic filtering capability of the MG, reduction in the compensation cost, and more flexibility of the distributed harmonic compensation schemes. It was also shown that ESS activities in harmonic compensation do not have much of an effect on the ESSs revenue from energy arbitrage. Hence, it could make ESSs more justifiable for use in MGs.

Keywords: energy storage system; active harmonic filter; harmonics; power quality; optimization

---

1. Introduction

Energy storage systems (ESSs) have received special attention due to their great flexibility and applicability in power systems. The smart charging and discharging behavior of ESSs could produce many advantages for various power system applications. The energy arbitrage for the operation of ESSs as consumers of low-price energy at off-peak periods and suppliers of high-price energy at peak hours is the most favorable application of ESSs. Furthermore, they could provide some advantages such as providing ancillary services and deferring investment costs in various sections of the power and distribution systems (DSs) [1–4]. All of these applications are related to the supply and security aspects of the energy. However, utilization of ESSs for the quality aspect of the energy for power quality (PQ) improvements has not gained much attention until now. With respect to the major costs that could be imposed by PQ related problems [5,6], it seems that investigations should be carried out regarding possible the PQ advantages of ESSs. Since the penetration of nonlinear power electronic (PE) based loads is increasing, it seems that ESSs could play an important role in enhancing PQ levels affected by these new loads in future DSs.

An ESS converts electrical energy via a power interface that is usually based on a PE converter. The general characteristics of ESSs are typically determined by their PE interface. Aside from the PE interface, various management and control components are included in ESSs, which give them various functionalities in the network. Flexibility and controllability of PE interfaces could be employed for PQ improvement in DSs. In this paper, PQ improvement using ESSs is discussed.

Any deviation that occurs from a pure sinusoidal voltage or current can be seen as a PQ problem. These can be listed as various issues such as harmonics, interruptions with different durations, and voltage related problems [7]. Problems related to the poor PQ levels impose remarkable costs on power systems [5,6]. Among these, harmonics are becoming more important due to an increase in the number of nonlinear harmonic loads and PE converters connected to the network, especially in DSs and microgrids (MGs). These loads could be employed in various applications from residential and commercial loads to industrial ones. Hence, harmonics could be imposed on the MG in any location at each voltage level. It has been reported that about 27% of all PQ problems are related to the harmonic distortions in voltage and current [8,9]. Harmonics could increase losses in DSs up to 20%. The losses are directly dependent on the number of nonlinear loads connected to the grid. A smart MG with a voltage THD ranging from 1% to 8%, could lose its generation capacity by 4.7% to 42.2%, respectively [5].

ESSs could play an important role in the mitigation of PQ related problems, especially regarding harmonics [2,4,7,10]. The ability to control PE converters could enable ESSs to act as active harmonic filters (AHFs). The change in the control algorithms of PE converters for taking additional services is discussed in the literature. The battery storage system is the most promising ESS in this scope. A smart battery controller was proposed in [11] for enhancing PQ and adjusting the steady-state voltage and frequency of MG. In [12], the optimal planning of ESSs for the mitigation of PQ problems was investigated to make ESSs more reasonable to be used in power systems. An investigation was made in [13] on the PQ of wind turbines with and without ESSs. The result shows the positive effects of ESSs as a solution for the mitigation of the harmonics of wind turbines. Battery energy storage could simultaneously compensate active, reactive, and harmonic power in four quadrants. A control approach for the compensation of voltage drop and harmonics was presented in [14]. In [15], the possible utilization of battery inverters for mitigation of harmonics and unbalances was reported. A harmonic and unbalance current controller was introduced based on the proportional-resonant controller in [15].

It should be noted that conventional control methods do not consider all other AHFs to calculate the compensation sharing of each AHF. In local controls such as current or voltage modes for eliminating current or voltage harmonics, AHFs work independently to other filters. Some control methods such as droop control provide some level of compensation sharing for a limited number of AHFs. However, the droop control is affected by the harmonic impedance of the network and is disabled to provide proper harmonic compensation sharing when AHFs become distributed. In this paper, an optimization-based model was proposed for achieving a coordinated operation of AHFs. In this approach, the MG operator tries to minimize the harmonic compensation cost with respect to the network constraints, working limits of AHFs, ESS activities in energy and ancillary markets, and harmonic standards. Hence, all constraints of AHF as well as the required harmonic compensation actions could be considered in the proposed method as a coordinated operation of AHFs.

It has been shown that optimal harmonic compensation can be achieved by the coordinated control of PE-based inverters [5,16]. This approach could avoid possible instabilities and ensure an optimal compensation scheme. Hence, the coordinated control of ESSs was implemented in this paper in order to show the capability of ESSs for distributed harmonic compensation. To this end, a new optimization model named as ESS-AHF was proposed for the calculation of the harmonic compensation sharing of each ESS. In this model, the MG operator tries to minimize harmonic compensation costs limited to various operational and technical constraints. The results of energy and ancillary service are used as setpoints for the ESSs to determine harmonic power truncations in each harmonic frequency. It should be noted that the distributed compensation scheme has better applicability to cope with the harmonics of the dispersed nonlinear loads of the MG. Furthermore, for reliable operation and unwanted stability problems, the coordinated operation of the AHFs should be ensured. Therefore, the proposed ESS-AHF model could successfully meet the requirements for distributed compensation and stability considerations.

The rest of this paper is organized as follows. In Section 2, active harmonic filters are introduced. Various ESS applications and characteristics are investigated in Section 3. A harmonic model of ESSs in harmonic compensation activities is also represented in this section. The proposed ESS-AHF model is introduced in Section 4. In Section 5, the results of the implementation of the introduced models are provided and analyzed. The paper then presents our concluding remarks in Section 6.

2. Active Harmonic Filters

Conventional mitigation methods of harmonics can be categorized as passive filters, static compensators, and active power line conditioners (APLCs). Currently, passive filters are the most prevalent technology for harmonic compensation. They consist of a single tune, second tune, and multi tune topologies. The passive filters are economic solutions and there is a great body of literature regarding their sizing and application for DSs [17]. Static compensators follow a principle similar to that of passive filters. However, these solutions have some limitations. Insufficient flexibility is the biggest disadvantage of passive solutions. Passive filters are usually designed for a specific harmonic condition, and hence change in the harmonic distortion or working conditions could degrade the applicability of the passive filters. Furthermore, resonance problems could affect the performance of passive filters and could increase harmonic distortion in some points of the network. Due to higher flexibility, there is an increasing approach in PE-based active harmonic filters (AHFs) such as APLCs. They could be optimal choices to cope with PQ related problems. In addition to the conventional APLCs, almost all energy resources with PE interfaces may be controlled to act, also, as harmonic active filters. Among them, ESSs equipped with PE converters could also be accounted for a possible AHF resource [18–21].

APLCs are considered as one of the most promising solutions for harmonic compensation activities. These devices do not suffer from conventional limitations of passive filters such as limited tuning, and especially resonance problems. They have the flexibility to work at a broad range of harmonics at any working point. Hence, APLCs are going to be more employed in electric networks [22–24]. An APLC is, in fact, a voltage or current source inverter that is controlled to produce harmonic compensation currents or voltages at the output. The APLC injects a harmonic compensation voltage or current to the network and eliminates harmonics in the source side. Unlike passive filters, APLC produces no resonance in the network. Hence, they are more prevalent, despite their higher investment costs and more complex control methods [22–24].

The schematics of two common types of APLCs called shunt and series topologies are represented in Figure 1. In both types, the voltage and current measurements of the network are processed by the APLC controller. The controller determines the required reference harmonic currents that should be injected by the filter. In [25,26], common control strategies for producing the desired reference current have been investigated. The calculated reference currents are then injected into the grid by a proper switching method. The required power for producing compensation currents can be supplied by an external source or taken by the grid using the same power converter.

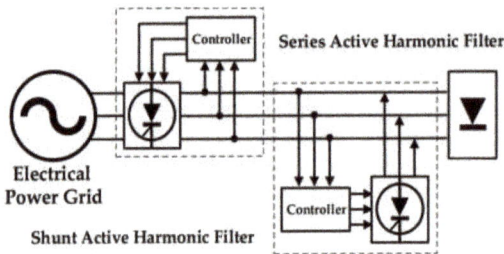

Figure 1. Schematics of the shunt and series active harmonic filters.

It has been shown that distributed compensation schemes could bring more flexibility and filtering capability for harmonic compensation [5,16]. Since the harmonic compensation is provided all over the grid, it is preferable to compensate current harmonics as close as possible to the disturbing source. In this way, harmonic losses can be reduced in the network. Harmonic loads are usually positioned at various locations with different nonlinear characteristics. Therefore, the distributed compensation scheme could have better applicability to cope with the harmonics of the dispersed nonlinear loads. Furthermore, for reliable operation of the MG, the coordinated operation of the AHFs should be ensured. It has been highlighted that instability problems can arise in the case of uncoordinated operation [5,16].

It should be noted that the penetration of nonlinear loads is increasing in MGs and compensation methods solely based on APLCs could be expensive and insufficient. Investment and operation costs are usually high for APLC devices [5]. Hence, other possible AHFs based on other PE converters should be utilized for more economic and flexible solutions. There is a great body of literature about the possible utilization of ESSs for PQ improvement [2,4,6,7,11,13,27]. However, to the best of the author's knowledge, the coordinated operation of ESSs for PQ problems has not been paid attention thus far. In this paper, the possibility of storage systems for the coordinated mitigation of harmonics is addressed. The coordinated operation of ESSs was achieved by an optimization problem including various constraints that could guarantee the economic and reliable operation of PQ improvement activities. In the next section, the possible utilization of ESSs for PQ improvement was investigated.

3. Energy Storage Systems for Power Quality Improvement

Energy storage systems have received special attention due to their great flexibility and applicability to the various power system sections. In MGs, an ESS converts electrical energy via a power interface, which is usually based on a PE converter. The general characteristics of ESSs are typically determined by their PE interfaces. Aside from PE interfaces, various management and control components are included in ESSs that enable them to have various functionalities in the network.

The main purpose of incorporating ESSs is to participate in energy markets. The term energy arbitrage refers to the charging of ESSs in the off-peak periods and discharging in the peak periods. The ESS revenues come from the difference between the off-peak and peak energy prices. Transmission and distribution services could also benefit from ESSs. The extension of network lines could be deferred using the flexibility of ESSs. This may avoid congestion problems in both the transmission and distribution grids, which result in more reliable operations of the system as well as reduced prices of energy [2]. Much work has been done on the integration of DERs in MGs. In MGs, the fluctuating output power of DERs in both grid-connected and islanded modes can cause voltage instability. This can also result in intermittent energy resources such as wind and solar farms. The ESSs could play a role for energy buffering and could smooth fluctuations of intermittent DER by a smart charging and discharging procedure. ESSs could then participate in voltage regulation and compensate reactive power to enhance the stability of the system [2].

One of the main applications of ESSs is in the area of ancillary services. Among various services, frequency regulation is the main service that could be supplied due to the fast response speed of PE interfaces. ESSs with high energy capacity could provide fast-spinning reserves. Furthermore, ESSs could act as AHFs and compensate for various PQ problems. Integrating ESSs in various DERs and flexible alternating current transmission system (FACTS) devices could enable this equipment for the compensation of extra reactive and harmonic currents. They could also avoid the PQ problems related to the application of DER in the MG.

ESSs vary in a broad range of technologies and categories. A common classification method is based on the form of stored energy. Mechanical (such as pumped hydro energy storage, compressed air energy storage (CAES) and flywheel), electrochemical (such as secondary batteries as the most useful technology), chemical (such as fuel cell), thermal (such as cryogenic energy storage and high-temperature thermal energy storage), and electrical (such as ultra-capacitors and superconducting energy storage systems (SMES)) are the main categories. Review papers about various ESSs technologies are available in [7,10]. In [10], the current practical implementations of ESSs all around the world were investigated.

Typical applications of ESSs are usually specified based on their properties in terms of power and energy capabilities. In [4], a comprehensive comparison was made among various ESSs in which output powers could be represented based on output duration, which is shown in Figure 2. Various battery types and mechanical energy storage systems such as pumped hydro and CAES with high energy density could be used in applications such a load-leveling and emergency power. Some other ESS types such as SMES and flywheels with conventional bearings have high power capacity. Hence, they are suitable for some PQ problems such as instantaneous voltage drop, voltage flicker, and short duration interruptions. Other ESSs such as flywheels with levitation bearing and super-capacitors can be employed for high power requests for a short time [4]. However, the duration depends on the requested power from the ESS.

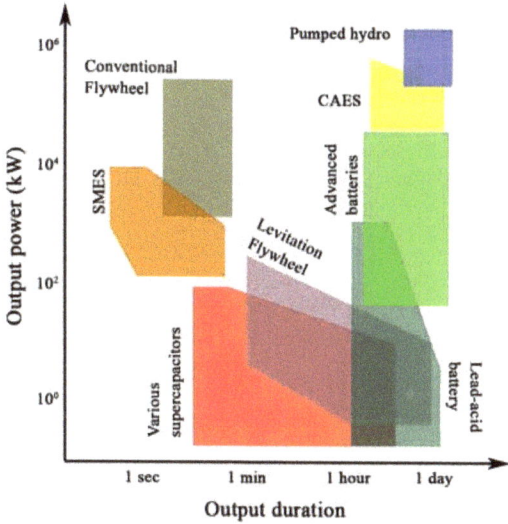

Figure 2. Output power and duration of various ESSs (plot obtained with data taken from [4]).

The general connection of some ESSs suitable for PQ mitigation activities is represented in Figure 3. The energy train may consist of a back-to-back converter for power conditioning between the MG and the main energy source. For example, flywheels work with AC power with their permanent magnet motors. Hence, they need an AC/DC converter for charging the DC link. Sometimes, a DC/DC converter can be employed before the DC link, as shown in Figure 1. After the DC link, a grid side converter (GSC) is used to deliver the energy from the DC link to the PCC. In order to enable the ESSs to participate in PQ mitigation activities, the control methods of the GSC should be modified.

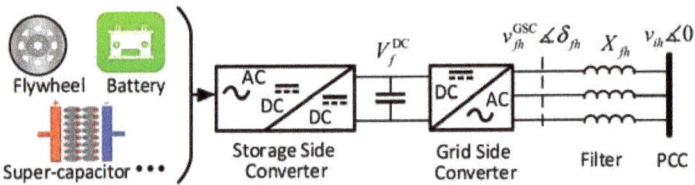

Figure 3. General connection of the ESSs by a voltage source inverter to the PCC.

Application of ESSs for Harmonic Compensation

Aside from AHFs and passive filters, ESSs could play an important role in the mitigation of PQ and harmonic problems [2,4,7,10]. The ability to control the PE converters of ESSs could enable them to act as harmonic active filters. In the literature, there are some examples of the modification of the control algorithms of PE converters for harmonic compensation. The battery storage system is currently the most promising ESS due to economic considerations and mature technology [27]. A smart battery controller was proposed in [11] for enhancing the PQ and adjusting steady-state voltage and frequency of MG. An investigation was made in [13] into the PQ of wind turbines with and without ESSs and the results showed the positive effects of ESSs as a solution for the mitigation of wind turbine harmonics.

Optimal harmonic compensation and the economic employment of ESSs could be achieved by the coordinated control of PE-based inverters [5]. It seems that the optimal operation of ESSs could efficiently consider various technical and systematic constraints in the calculation harmonic sharing of each ESS. Hence, the coordinated control of ESSs was proposed in this paper for harmonic filtering activities as an extension for various applications of ESSs. Therefore, the utilization of various ESSs for harmonic compensation and the coordinated control of ESSs for a global distributed harmonic compensation scheme were proposed and formulated. To this end, a harmonic model of ESSs is required in the coordination algorithm. Based on Figure 3, the equivalent circuit model of ESSs in harmonic mitigation can be represented as in Figure 4. The internal current source of the ESS is controlled based on the internal parameters of the ESS such as the voltage of the DC link, control, and switching method of the inverter. The GSC is controlled in the current control mode for harmonic compensation [27].

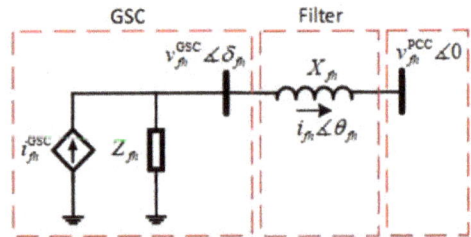

Figure 4. Harmonic model of ESSs in harmonic compensation.

Internal impedance Z_f shows the Thevenin impedance of the ESS seen in the output of the GSC. The internal impedance depends on the internal parameters, the control blocks, and the topology of the GSC. Some calculation methodologies of the GSC internal impedance as a VSI was investigated in [28–30]. A simple model for the Z_f can be represented as Equation (1), which includes two resistive (R_{fh}) and inductive parts (X_{fh}). Both parts are considered as functions of various internal parameters such as the voltage of the DC bus (V^{DC}), internal inductance (L), and also as gain factors of the internal control blocks (G_{fh}) of the GSC. The current gain, the voltage gain, the gain of current control compensator, the feedforward gain, and the modulator gain are internal control parameters affecting G_{fh}, which should be considered in the calculation of Z_{fh} [28].

$$Z_{fh}^{GSC} = R_{fh}^{GSC}\left(G_{fh}^{GSC}, V_f^{DC}, L_f\right) + jX_{fh}^{GSC}\left(G_{fh}^{GSC}, V_f^{DC}, L_f\right) \tag{1}$$

Finally, it should be noted that the reactance of X_{fh} was employed for the output filter of the ESSs, as shown in Figure 3. The proposed model in Figure 4 could be employed for the coordination of the distributed ESSs for harmonic compensation. In the next section, the proposed model for harmonic compensation with ESSs is introduced.

4. Coordination Model of ESSs for Harmonic Filtering

In this section, the proposed model for the coordination of ESSs for harmonic mitigation is introduced. The model was developed based on the concept of the harmonic power flow [5,31] and active power filter sizing and placement [32]. Since the main power transactions of ESSs are in the fundamental frequency, the side operation for harmonic compensation could be implemented as a sequential operation model. Hence, the harmonic spectrum approach could be employed to calculate the harmonic filtering sharing of each ESS after running the day-ahead energy market [33]. In the following, the optimization model for incorporating ESSs in the operation of the MG at the fundamental frequency is first presented. After that, the proposed model for the coordination of ESSs working as AHFs (named the ESS-AHF model) and related working constraints are introduced.

4.1. Operation in Fundamental Frequency

In the fundamental frequency, ESSs can be used in the daily operation of the MG for energy arbitrage. This problem can be formulated as an operation problem including ESSs [34,35]. The goal of the MG operator is to economically supply the load demand using energy purchased from the main grid ($cost^{grid}$), the energy of the available DGs ($cost^{DG}$), and eventually, the energy arbitrage of ESSs ($cost^{ESS}$). Hence, the operation cost (OC) can be mathematically formulated as the minimization problem shown in Equation (2).

$$\min\left\{OC = \sum_t \left(cost_t^{grid} + cost_t^{DG} + cost_t^{ESS}\right)\right\} \tag{2}$$

The objective function in Equation (2) consists of three main costs. The energy purchased from the main grid can be calculated with respect to the price of energy and power imported from the main grid to the MG as shown in Equation (3). Furthermore, DGs could also inject power to the MG whose related costs can be calculated as per Equation (4) in each operation period. For ESSs, the costs should be calculated with respect to the injected power in discharging mode and consumed power in charging mode. This can be mathematically formulated as in Equation (5). In Equations (3)–(5), the terms of π stand for prices of energy in the main grid for the injected power of DGs and prices for the charging and discharging of ESSs, respectively.

$$cost_t^{grid} = \pi_t^{grid} p_t^{MG} \tag{3}$$

$$cost_t^{DG} = \sum_i \left(\pi_{it}^{DG} p_{it}^{DG}\right), \forall i \in \Omega^{DG} \tag{4}$$

$$cost_t^{ESS} = \sum_i \left(\pi_{it}^{dch} p_{it}^{dch} - \pi_{it}^{cha} p_{it}^{cha}\right), \forall i \in \Omega^{ESS} \tag{5}$$

There are some technical and systematic constraints for the minimization of the OC, which should be taken into account. For ESSs, the working constraints are shown in Equations (6)–(10). Three possible working modes of ESSs can be considered: charging, discharging, and idleness modes.

These modes can be modeled as Equation (6) by using the binary variables of y^{cha} (stands for charging mode) and y^{dch} (stands for discharging mode). If these two variables are simultaneously adjusted to zero, then the ESS is idle in that period. The state of charge (SOC) of ESS (c_{et}) was calculated in Equation (7) in each operation period. It was determined based on the SOC at the beginning of the period in addition to power transactions in charging (p^{cha}) or discharging (p^{dch}) modes. In Equation (7), charging and discharging efficiencies are also considered as η^{cha} and η^{dch}, respectively. Constraint (8) is proposed for the calculation of the injection current (i) of the ESS to the connection bus in each mode using the conjugate of bus voltage, v_{it}, and the active power of ESS. Since the operation modes are separated using binary variables in Equation (6), p^{cha} and p^{dch} do no occur simultaneously. Hence, in Equation (8), either charging or discharging of the ESS could take place. Consequently, the injection current could be positive or negative, with the injection current considered to be positive in charging mode. In Equation (9), the working limits of the SOC are considered using C^{min} and C^{max}, respectively, for the minimum and maximum limits. The maximum and minimum limits of active power are also considered in Equation (10).

$$y_{et}^{cha} + y_{et}^{dch} \leq 1 \tag{6}$$

$$c_{e,(t+1)} = c_{et} + \eta_{le}^{cha} p_{et}^{cha} - \frac{1}{\eta_{le}^{dch}} p_{et}^{dch} \tag{7}$$

$$i_{it} = \frac{p_{et}^{cha}}{v_{it}^*} - \frac{p_{et}^{dch}}{v_{it}^*}, \forall i \in \Omega^{ESS} \tag{8}$$

$$C_e^{min} \leq c_{et} \leq C_e^{max} \tag{9}$$

$$P_e^{min} \leq p_{et}^{cha}, p_{et}^{dch} \leq P_e^{max} \tag{10}$$

Power flow constraints are the main system constraints in the optimization model. Current injections of loads, DGs, and nonlinear loads could be calculated in Equation (11) using the apparent power (S) and the voltage of the connecting bus (v_i) at each operation period. The injection shown in Equation (11) is valid for the set of load buses (Ω^{PQ}), the set of DG buses (Ω^{DG}), and finally the set of nonlinear buses (Ω^{NL}). Without losing generality, it is assumed that the DGs work in constant power mode. Anyway, other operation modes could simply be applied in the model [31]. Furthermore, the efficient linear power flow constraints of [31] were employed for modeling the network. In this power flow model, the MG was modeled with some incidence matrices using graph theory. Power flow constraints are modeled in Equations (12) and (13) using A and B incidence matrices employed for modeling the MG. In these equations, the matrices of the line currents, bus injections, and voltage drops of buses are shown with $[U]$, $[I]$, and $[\Delta V]$, respectively. It has been shown in [31] that this power flow model has no convergence problem in MGs and can easily handle meshed topologies and unbalanced networks.

$$i_{it} = \left(\frac{S}{v_{it}}\right)^*, \forall i \in \{\Omega^{PQ}, \Omega^{DG}, \Omega^{NL}\} \tag{11}$$

$$[U] = [A][I] \tag{12}$$

$$[\Delta V] = [B]^T [Z][U] \tag{13}$$

Nevertheless, modeling the energy arbitrage mechanism of ESSs was not the main concern of this paper. The output of the fundamental frequency optimization model is the SOC of each ESS and the working points in the operation horizon were used for calculations in harmonic frequencies.

4.2. Coordinated Operation of ESSs as AHF

In this subsection, the proposed model for employing ESSs as AHFs (named the ESS-AHF model) is introduced. The goal of the MG operator is to minimize the cost of procurement of PQ improvement actions. Since the harmonics were only investigated in this paper, this goal could be reached using

AHF resources, APLCs, and ESSs. Hence, the objective function can be mathematically shown as Equation (14).

$$\min\left\{PQ\ cost = \sum_{th}\left(cost_{th}^{APLC} + cost_{th}^{ESS}\right) = \sum_{fth}\left(\pi_{fh}^{APLC}d_{fth}^{APLC} + \pi_{fh}^{ESS}d_{fth}^{ESS}\right)\right\} \quad (14)$$

The cost of PQ improvement actions includes two main costs: costs related to APLCs denoted by the $cost^{APLC}$ and costs related to the ESSs working as AHFs, shown with $cost^{ESS}$. The calculation of these two cost terms is also represented in Equation (14) using the amount of the provided compensation distortion power (d) and the offered compensation price (π) for each AHF (f) at each period. In harmonic frequencies, due to the increase in power losses of the AHF, some amount of distortion power is imposed on the AHF. The payments are required for covering such costs for the owners of AHFs. Hence, offered prices of harmonic compensation, currents should be capable to refund extra costs imposed on the filter. On the other hand, since the required compensation actions are supplied in a competitive mechanism, the offered prices should ensure sufficient participation of the ESS owner in PQ improvement activities. Moreover, the offered prices of ESSs are generally lower than the prices of APLCs, since the harmonic compensation action is taken as a side function from ESSs [5]. Hence, some operation costs related to the investment cost can be decreased for ESSs. Anyway, the cost of all AHFs can be calculated using the offered price of each AHF and the related provided compensation distortion power as shown in Equation (14).

The distortion power generally includes the current distortion power [Marini, 2019 #445]. Hence, the provided compensation distortion power could be shown as Equation (15) using the provided harmonic compensation currents in each harmonic order, i_{fth}. In this equation, the distortion power is assumed to be limited by the maximum available power of the AHF, D^{max}. For APLCs that do not contribute at the fundamental frequency, this limitation is simply considered as the rating volt-ampere of the filter. However, for AHFs based on ESSs, considerations should be made for power transactions in the fundamental frequency. This limitation is set as the maximum value of available power of the ESS ($P^{max}-P_{et,1}$) for idle periods and allowable 10% variation in output power ($P_{et,1}$) for charging or discharging periods. This constraint is mathematically shown as the maximum value of these two values in Equation (16). Furthermore, with respect to the general model of AHF as shown in Figure 4, working constraints could be formulated as Equations (17) and (18). In harmonic studies, harmonic voltages and currents are superimposed to the fundamental voltage and current. Hence, voltage and currents should be replaced with harmonic root mean square equivalents, as shown in Equations (17) and (18). The maximum current of AHF is limited to the maximum value, I^{max} in Equation (17). In Equation (18), the maximum allowable limit of AHF is considered as V^{max}.

$$d_{ft}^2 = V_{ft,1}^2 \sum_{h>1} i_{fht}^2 \leq \left(D_{ft}^{max}\right)^2 \quad (15)$$

$$D_{ft}^{max} = \max\left\{P_e^{max} - P_{et,1}, 10\%(P_{et,1})\right\}, \forall f \in ESS \quad (16)$$

$$I_{ft,1}^2 + \sum_{h>1} i_{fht}^2 \leq \left(I_f^{max}\right)^2, \forall f \quad (17)$$

$$V_{ft,1}^2 + \sum_{h>1} v_{fht}^2 \leq \left(V_f^{max}\right)^2, \forall f \quad (18)$$

The SOC of the ESS should also be taken into account in harmonic conditions. Although harmonic power transactions have low values, their effects on the SOC of the ESS should be considered. Harmonic active power (p_{fth}) can be calculated using the real value of the product of harmonic voltage and current as shown in Equation (19). The SOC of the ESS should be modified with respect to the provided active harmonic power. Since the PE converter of the ESS is generally a four-quadrant converter, harmonic active powers could be positive and negative. This could lead the ESS to be charged or discharged to provide the harmonic compensation current. Variations in the SOC in harmonic condition (c_{fth}) with respect to the provided active harming powers is mathematically shown in Equation (20). In Equation (20), the SOC in the harmonic condition is supposed to be equal to the SOC for fundamental frequency (C_{ft}) plus harmonic active powers, p_{fth}. Harmonic reactive power is usually achieved by changing the phase angle and modulation method of the converter. It can be considered as extra losses of active power and is usually considered as a percentage of active harmonic power [36]. This modification could be applied in Equation (20), which could be a subject for future research. Finally, the minimum and maximum limitations of the SOC should also be met in harmonic activities, which is mathematically shown in Equation (21).

$$p_{fth} = \text{Re}\{v_{ith} i_{ith}^*\}, \forall i \in \Omega^{ESS} \tag{19}$$

$$c_{fth} = C_{ft} + \sum_{h>1} p_{fth}, \forall f \in \text{ESS} \tag{20}$$

$$C_f^{min} \leq c_{fth} \leq C_f^{max}, \forall f \in \text{ESS} \tag{21}$$

Harmonic standards are usually reported as total harmonic distortion (THD) and individual harmonic distortion (IHD). Typical levels of harmonic standards for grid voltage are 5% and 3% for THD and IHD, respectively [37]. The important goal of the MG operator is the economic satisfaction of the desired harmonic levels across the entirety of the MG at each period. These constraints are shown in Equations (22)–(23), respectively, for THD and IHD. In each operation period, the THD could be met for each bus and the IHD should be met for each bus in each harmonic order.

$$\sum_{h>1} v_{ith}^2 \leq |V_{ih,1}|^2 (\text{THD}_i^{max})^2, \forall i, t \tag{22}$$

$$v_{ith}^2 \leq |V_{it,1}|^2 (\text{IHD}_i^{max})^2, \forall i, \forall t, \forall h \tag{23}$$

The harmonic power flow constraints should also be considered in the model. Since the provided harmonic current compensation is injected to the MG in different nodes, these constraints are essential for calculations of the flow of harmonic powers. The constraints of Equations (24) and (25) are the harmonic counterparts of fundamental power flow previously shown in Equations (12) and (13). Furthermore, the harmonic spectrum approach is used for the harmonic modeling of nonlinear loads [33]. It is required to determine the magnitude and phase angle of harmonic injections of nonlinear loads in this approach. In Equation (26), the magnitude of harmonic injection (i_{iht}) is determined based on the harmonic spectrum of nonlinear load (I^{spec}) and current injection of the load in the fundamental frequency ($I_{it,1}$) for all buses of the network with nonlinear load (Ω^{NL}). The phase angle of harmonic injection (θ_{ith}) was also determined in Equation (27) using the phase spectrum of the nonlinear load (θ^{spec}) and the phase angle of its current injection in the fundamental frequency ($\theta_{it,1}$).

$$[U_h] = [A_h][I_h] \tag{24}$$

$$[\Delta V_h] = [B_h]^T [Z_h][U_h] \tag{25}$$

$$i_{iht} = I_{iht}^{spec} \times I_{it,1}, \ i \in \Omega^{NL} \tag{26}$$

$$\theta_{ith} = \theta_{ith}^{spec} + h\theta_{it,1} + \pi, \ i \in \Omega^{NL} \tag{27}$$

The developed model in Equations (14)–(27) is the complete model proposed for the coordinated operation of ESSs working as AHFs. It should be noted that the operation model of Equations (2)–(13) and the ESS-AHF model of Equations (14)–(27) are related and the ESS-AHF model will be implemented after the operation model. Hence, the ESS-AHF model could be accounted for as a sequential model, as shown in Figure 5. The output nonlinear load current of the operation model is the setpoints for the calculation of harmonic injections. These dependencies are mathematically shown in Equations

(26) and (27). Harmonic injections are calculated in Equations (26) and (27) based on the harmonic spectrum of the nonlinear load. Hence, these dependencies could affect the ESS-AHF model since the harmonics will be related to harmonic injections of nonlinear loads. Furthermore, since ESSs are employed as AHFs in addition to APLCs, the working conditions of ESSs could be affected by the output of the operation model. The maximum available capacity of the distortion power for ESSs is set as the maximum value of available power of the ESS and allowable 10% variation in output power for charging or discharging periods, as shown in Equation (16). Hence, these setpoints could change the ESS activities as AHFs.

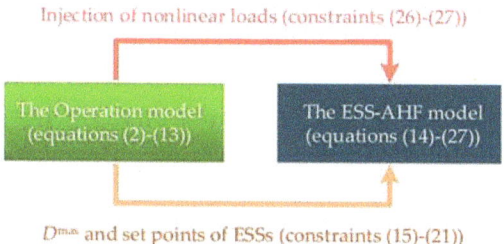

Figure 5. The ESS-AHF model as a sequential model after the operation model.

The output of the ESS-AHF model is the calculated harmonic reference currents of the ESSs, which should be injected into the MG. These reference currents are then transmitted to the ESSs by the available infrastructures of the MG, and the AHF works in the current control mode. Various technical constraints, harmonic conditions of the network, and the status of AHFs should be sent to the MG control center for consideration in the optimization problem. All of these require a proper communication infrastructure that could be available in future MGs. The availability of communication links in AHF buses, its bandwidth, and time delay are important challenges of the communication system that could affect the applicability of the proposed optimization method, which should be taken into account in future studies. In the next section, the results of the implementation of the proposed model will be represented.

5. Simulations

A modified version of the IEEE 33-bus test network was used to demonstrate the performance of the ESS-AHF model. The test grid consists of 32 branches and five tie-switches, as shown in Figure 6. Line and bus data were taken from [38] and are provided in Appendix A. In order to increase the effects of harmonics, line impedances were increased by a factor of three. Data for the daily load demand were taken from [39].

Four nonlinear loads were connected to the test network. The locations and active and reactive powers of nonlinear loads are shown in Table 1. A general non-linear load model as a six-pulse diode bridge rectifier was adopted for all nonlinear loads [33]. The represented harmonic loads are the main sources of harmonic distortions in the MG. The harmonic spectrum of nonlinear loads is reported in Table 2, where the data were taken from [33]. In the harmonic spectrum approach, the magnitude and phase of the harmonic injections of nonlinear loads are used to calculate the flow of harmonic power in the network individually at each harmonic frequency.

Five DERs were added to the test network in selected buses, as shown in Figure 6. A parking lot of electric vehicles was considered at bus #12, a wind turbine was installed at bus #21, a photovoltaic (PV) generator was installed at bus #24, a battery energy storage was connected at bus #28, and finally, a PV array was connected at bus #31 equipped with a battery energy storage. The parking lot at bus #12 included a battery ESS. Hence, three ESSs have been placed in the test network which could be used as possible AHFs. Furthermore, two APLCs were connected to the network in buses #2 and #9, as shown in Figure 6. Hence, the total number of AHFs was equal to five filters. Data of the DER and

APLCs are presented in Table 3. The maximum and minimum of the SOC and maximum power of each ESS are also reported in this table. In the last column, the offered price of distortion power is shown based on $/kVARh. The offered prices for ESSs were lower than the APLCs since harmonic filtering is considered as a side function from ESSs, which leads to a reduced distortion power cost.

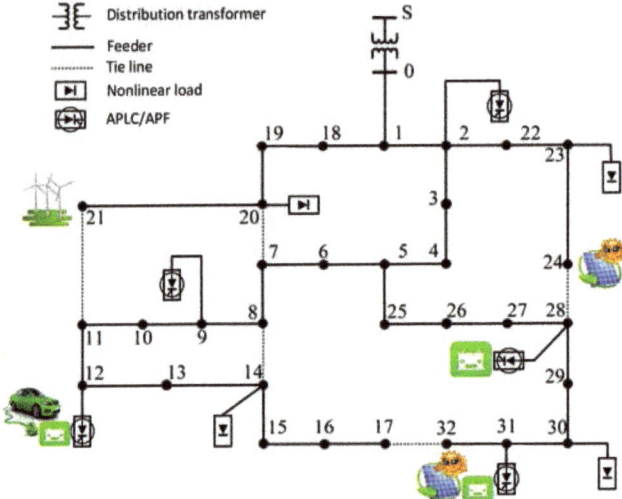

Figure 6. Modified IEEE 33-bus test network.

Table 1. Nonlinear loads of the network.

Nonlinear Load	Bus	P (pu)	Q (pu)
1	14	0.150	0.120
2	20	0.125	0.105
3	23	0.110	0.100
4	30	0.145	0.125

Table 2. Harmonic spectrum of nonlinear loads.

Harmonic	Magnitude	Phase (°)
5	0.35	180
7	0.43	180
11	0.05	0
13	0.08	0
17	0.04	180
19	0.04	180

The lower and upper bounds of voltage were taken as 0.95 pu and 1.05 pu, respectively. The operation horizon for unit commitment problem is considered the next day, which is divided into hourly periods. The proposed optimization models (in the form of a mixed-integer nonlinear programming model) were implemented in GAMS software (GAMS Development Corporation, Washington, D.C., USA) [40] in a personal computer equipped with a 2.93 GHz Intel processor and 8 GB memory. Without loss of generality, it was assumed that the MG works in grid-connected mode. Hence, the voltage at the PCC is assumed to be sinusoidal.

Table 3. Data of APLCs and DER of the network (values are in pu).

No	ID	Type	Bus	S^{max}	I^{max}	C^{min}	C^{max}	P^{max}	π ($/kVARh)
1	APLC #1	APLC	2	0.22	0.43	—	—	0.32	860
2	APLC #2	APLC	9	0.18	0.35	—	—	0.26	900
3	ESS #1	Electric vehicle + ESS	12	0.1	0.1	0.1	0.25	0.1	395
4	DG #1	Wind turbine	21	0.15	0.15	—	—	0.15	—
5	DG #2	Photovoltaic	24	0.08	0.1	—	—	0.08	—
6	ESS #2	Battery storage	28	0.15	0.25	0.15	0.50	0.15	380
7	ESS #3	Photovoltaic + ESS	31	0.1	0.35	0.1	0.22	0.1	380

Three different cases were considered to show the performance of the ESS-AHF model.

- Case I: ESSs are only used for energy arbitrage to reduce the cost of supplying energy to the customers.
- Case II: Employing APLCs for harmonic compensation imposed by nonlinear loads.
- Case III: Employing ESSs as AHFs to satisfy harmonic standards.

The simulation results are reported in the following three subsections.

5.1. Case I: The Base Case

In the base case, the unit commitment model of 4.1 was implemented on the test network. The various cost terms after running the model are summarized in Table 4. The total OC was equal to 2410 k$, 87% of which was due to the energy supplied by the main grid. The other 13% of the OC was due to the energy produced or exchanged by DGs and ESSs. The cost of the energy supplied by DGs was equal to 248 k$ and represented the revenue of DG owners for participating in the energy market. In the revenue of DG #2, the PV at bus #24 was higher compared to other DERs for its proper location. Moreover, smart charging and discharging of ESSs implied a revenue of 58 k$ from energy arbitrage. This revenue is, of course, a cost for the MG operator. In particular, ESS #2, the battery storage at bus #28, earnt about 50% of the total ESSs revenues.

Table 4. Operation cost for the test network in Case I.

Total OC (k$)	Grid Cost (k$)	DG Cost (k$)		ESS Cost (k$)		
		248		58		
2410	2104	DG #1	DG #2	ESS #1	ESS #2	ESS #3
		86.4	161.6	17	28	13

The charging profile as well as the SOC of ESSs in the operation horizon is shown in Figure 7. All ESSs were charged from 3 am to 6 am, which are the off-peak periods of the operation horizon. Similarly, ESSs are discharged in peak periods from 5 pm to 7 pm. In Figure 7b, the SOCs of ESSs are represented. Allowable limits in Table 4 for the SOC and maximum power are met by the optimization model. Variation of the SOC for ESS #2 with more energy capacity at off-peak and peak periods can be seen in this figure.

However, the harmonic conditions should also be considered for PQ monitoring of the MG. In Figure 8, the voltage THD of the MG was analyzed. The voltage THD of all buses at 5 am (as the lowest) and 6 pm (as the highest load demand), are shown in Figure 8a. Obviously, either for off-peak periods or peak periods, the THD was above the standard level for nearly all system buses. All other periods may have a THD trend between these two selected periods. Bus #17 had the maximum THD compared to other buses. In Figure 8b, the voltage THD of this bus was represented in the operation horizon. The standard THD level is never met and hence, compensation actions are necessary.

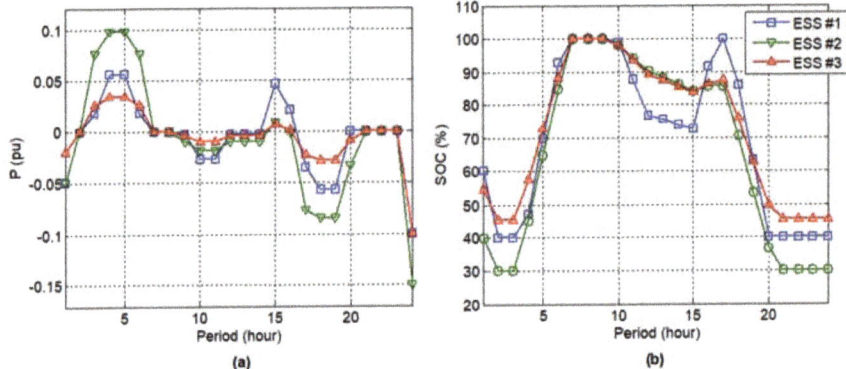

Figure 7. Case I. (a) Charging profile and (b) the SOC of each ESS.

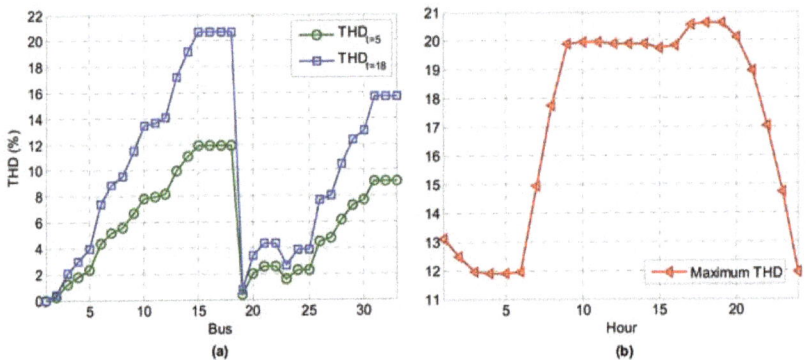

Figure 8. Case I. (a) Voltage THD for all buses at $t = 5$ and $t = 18$ and (b) maximum voltage THD of all buses at each hour.

In order to mitigate the harmonics effects from the grid, two simulation cases are provided to show the performance of the ESS-AHF model.

5.2. Case II: Harmonic Compensation Using APLCs

In Case II, only APLCs were employed for harmonic mitigation. APLCs were placed in the network based on filtering requirements and economic conditions by optimization models. The voltage THD of the system buses and the maximum THD after compensation of the APLCs are shown in Figure 9 for this case. Although the harmonic pollution was improved when compared to Case I, it was still above the standard level of 5%. This situation is worthwhile in peak periods, as shown in Figure 9a, since harmonic injections of nonlinear loads are increased in these periods. The maximum THD was for bus #32, which was about 11%. In Figure 9b, the maximum THD is shown for this bus. For off-peak periods, APLCs could adjust the voltage THD at the standard level. However, most of the time, the voltage THD is high, which could be the result of insufficient harmonic filtering capacity in the MG.

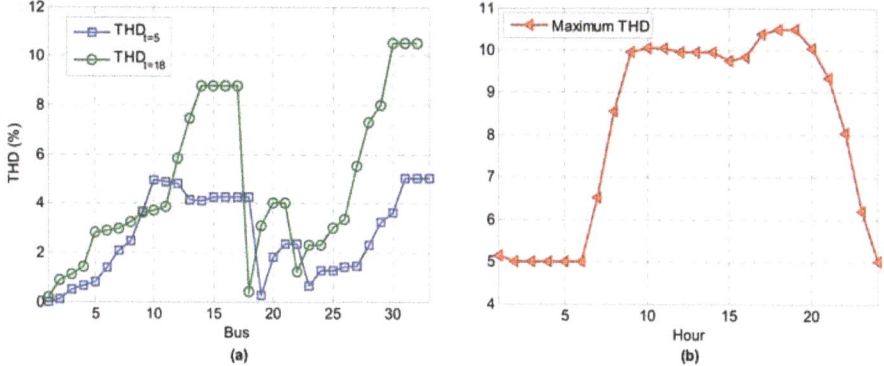

Figure 9. Case II. (a) Voltage THD for all buses at $t = 5$ and $t = 18$ and (b) maximum voltage THD of all buses at each hour.

In Figure 10 the loading of each APLC is shown for Case II. In almost all periods (except off-peak ones), the APLCs were loaded at their maximum rating. However, they could not satisfy the harmonic standard in the MG. Hence, more harmonic filters are required to be employed for more filtering capacity in the MG. This could be achieved using more APLCs. However, it seems that this is not an economic solution. With respect to the high investment costs of the APLCs, it could significantly increase the cost of PQ compensation actions.

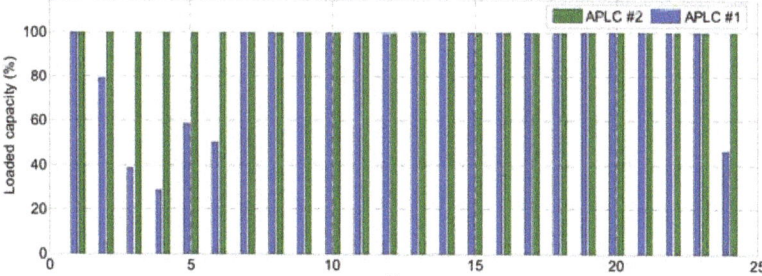

Figure 10. The loaded capacity of the APLCs for harmonic filtering in Case II.

In the next case, the performance of ESSs for improving the harmonic situation of the MG as an economical alternative to APLCs is presented.

5.3. Case III: Harmonic Compensation Using ESSs

In order to increase the harmonic filtering capability of the MG as well as reduce the costs of PQ improvement actions, ESSs are also employed in this case for harmonic compensation. To this end, the unloaded power capacity of ESSs is used for harmonic filtering activities. Even if the ESS does not contribute to energy arbitrage, it can provide harmonic filtering capacity for the network. The results of the implementation of the ESS-AHF model are represented in this subsection.

The voltage THD and the maximum THD at each hour are represented in Figure 11. As it could be seen, the maximum THD in all periods of the operation horizon was adjusted at the standard level. For either the off-peak periods or peak periods, the voltage THD was in the acceptable range. Hence, using ESSs as AHFs has provided more harmonic filtering capacity for the MG. Furthermore, since the ESSs are dispersed in the MG, the provided compensation currents could be considered as distributed harmonic compensation actions. In this scheme, harmonic compensation currents are injected in the grid, and hence harmonic propagation and the resulting losses are strongly reduced. In Figure 13b, the maximum THD of voltage buses is shown, which was adjusted to the 5% standard level. This shows the effects of ESSs for the adjustment of harmonic standards for MGs.

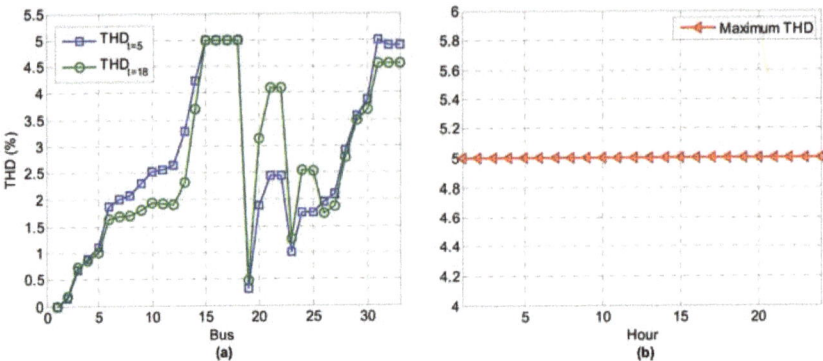

Figure 11. Case III. (a) Voltage THD for all buses at $t = 5$ and $t = 18$ and (b) maximum voltage THD of all buses at each hour.

The loadings of various AHFs are shown in Figure 12 for Case III. AHFs based on ESSs are almost loaded at their maximum capacity. This is the result of their lower prices for compensation distortion power as well as their location. Since the APLCs have expensive distortion power, their compensation loadings are performed after ESSs and generally have lower values compared to ESSs. APLC #2 had more compensation loading despite its higher cost. This shows the effects of filter location on its provided harmonic compensation to the MG.

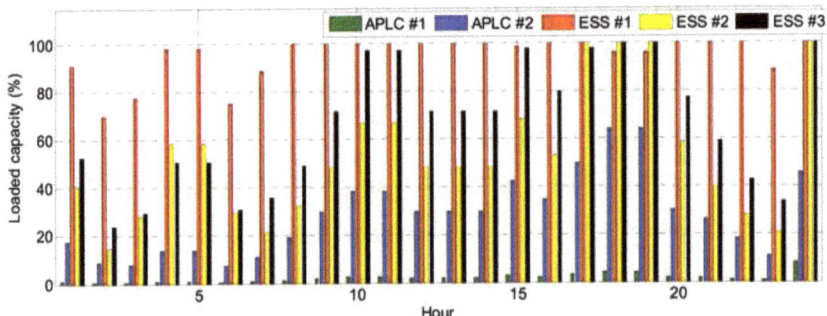

Figure 12. The loaded capacity of AHFs for harmonic filtering in Case III.

The harmonic active power and the SOC of ESSs after harmonic compensation activities are reported in Figure 13. The provided harmonic active power was negative for all ESSs, which shows that it worked in discharging mode for harmonic compensation. However, the harmonic injection powers had low values compared to the fundamental active power previously shown in Figure 7. This means that participating in harmonic compensation activities does not have a significant effect on the SOC of ESSs, which can be seen Figure 13b. The trend of the SOC was close to the trend of that in Figure 7 and all allowable limits were met in the harmonic conditions. Hence, the harmonic utilization of ESSs could bring various advantages for the MG, and meanwhile, does not affect the ESS parameters for energy arbitrage. The ESSs earn from both energy arbitrage and harmonic compensation.

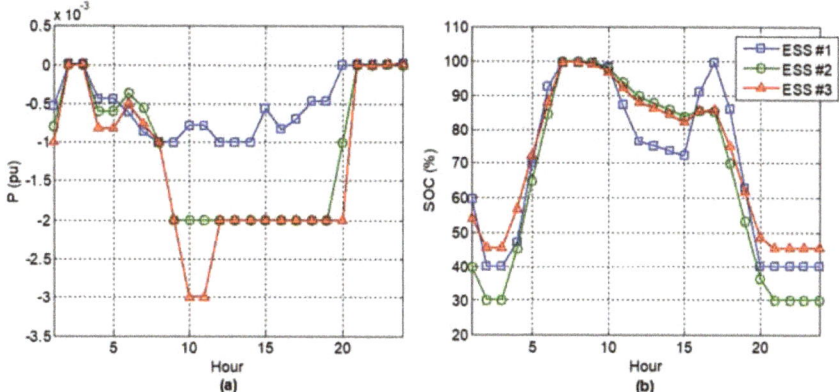

Figure 13. Case III. (a) Discharging profile and (b) the SOC of each ESS for harmonic compensation.

The required reference currents of the ESSs at each period are determined in the ESS-AHF model. Fundamental harmonic and reference currents for ESS #2 at the peak period (i.e., 6 pm) are shown in Figure 14. The bold curve is the required reference current calculated by the ESS-AHF model, which should be injected by the ESS. The reference current should then be sent to the ESS by the available infrastructures of the MG. The ESS is responsible to inject the required current by the MG operator with a suitable control and switching method.

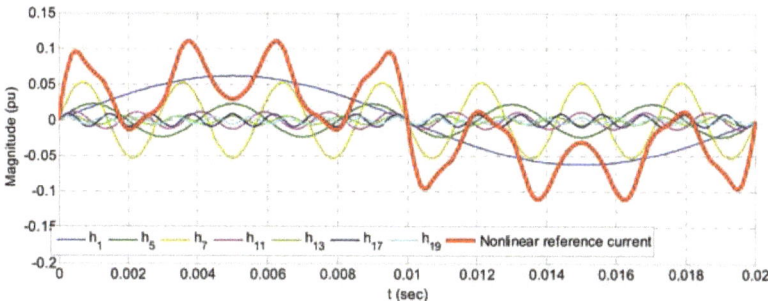

Figure 14. Fundamental harmonic and reference currents of ESS #2 in the peak period, (6 pm) in Case III.

For economic comparisons, the total injections and compensation costs for Case III are presented in Table 5. The total required distortion power was 8.95 pu and 6.29 for Case II and Case III, respectively. Although the compensation distortion power in Case III was about 70% of that in Case II, it could successfully meet the desired level of harmonics in the MG, as shown in Table 5. Furthermore, the loading of compensation currents was more achieved by the ESSs in Case III as they were more economic AHFs compared to the APLCs. This reduces the total compensation cost from 7868 k$ in Case II to 3424 k$ in Case II. It showed a 44% decrease in the compensation cost. Hence, the MG operator could keep the desired harmonic level with only 44% of the cost of the APLCs. This could save remarkable costs for the MG operator, either in the operation of the APLCs or as requirements for the installation of new APLCs.

Table 5. Total harmonic compensation injection of AHF in Case II and Case III.

	ID	APLC #1	APLC #2	ESS #1	ESS #2	ESS #3	Total
Case II	Injection (pu)	4.63	4.32	0	0	0	8.95
	Cost ($)	3980	3888	0	0	0	7868
Case III	Injection (pu)	0.16	1.79	1.74	1.29	1.31	6.29
	Cost ($)	140	1607	689	491	497	3424

The simulations showed the performance of the ESS-AHF model in the coordination of distributed ESSs for harmonic compensation in the MG. More filtering capacity and reduced compensation costs are the main advantages of ESSs in harmonic compensation. Since harmonic compensation is utilized as a side function, the provided distortion power has a lower price than APLCs. This implies that compensation loading is first demanded by the ESSs. Furthermore, ESSs are dispersed in the MG, and hence harmonic compensation will be distributed in the MG. The provided compensation scheme will be accounted for by distributed harmonic compensation with more flexibility and applicability for PQ improvements. The ESS owners could also benefit from both energy arbitrage and harmonic compensation. This could decrease the investment costs of the ESSs and make them more justified for use in the MG. The simulations showed that ESS compensation activities do not have much of an effect on the SOC and ESS revenues from energy arbitrage. Finally, it should be noted that, unlike energy and ancillary markets, PQ improvement actions are not security obligations of the network. Therefore, any time that the working of ESSs in PQ improvement actions affects their application in ancillary services and the security of the network, it could be temporarily ignored.

6. Conclusions

In this paper, the applicability of ESSs for harmonic filtering activities was investigated. The common procedure for this was to modify the control methods of the PE converters of ESSs to enable them for the injection of compensation currents. However, in this paper, the coordinated operation of ESSs for a distributed harmonic compensation scheme was addressed. The proposed coordination model employed conventional control algorithms for the injection of a reference current that was determined in the proposed ESS-AHF optimization model. Hence, the provided compensation scheme could be considered as a distribution compensation scheme. The simulation cases showed the performance of the proposed model in enhancing the filtering capability of the MG, the reduction in compensation cost, and the greater flexibility of the distributed compensation schemes. Hence, requirements for new filters could be deferred and both the MG operator and the ESS owner could benefit from reduced compensation costs.

ESSs can be used to perform several ancillary services where most of them need the energy stored to be accomplished. What is demonstrated in this paper is that the ancillary service for harmonic power filtering can be performed without significantly affecting the battery SOC. For this reason, this service can be achieved *for free*, without compromising the others, and therefore the revenues from energy arbitrage and other ancillary services can be maintained. This could make ESSs more economical for use in future power systems. Investigations should be carried out for various technical and systematic challenges of the proposed ESS-AHF model in future research.

Author Contributions: A.M. and L.P. conceived and designed the models; A.M. performed the experiments and simulation; A.M., L.P., and S.-S.M. analyzed the data; M.-S.G. contributed reagents/materials/analysis tools; A.M. and L.P. wrote the paper. All authors have read and agreed to the published version of the manuscript.

Funding: This research received no external funding.

Conflicts of Interest: The authors declare no conflicts of interest. The founding sponsors had no role in the design of the study; in the collection, analyses, or interpretation of data; in the writing of the manuscript, and in the decision to publish the results.

Nomenclature

Note that single parameters, single variables, and matrices are shown with capital letters, small letters, and capital letters in brackets, respectively, in the paper. When each notation comes with a specific index and superscript, it stands for the corresponding variable or parameter of related indices. Repetitive definitions can be ignored using this method. For example, the letter 'v' always stands for voltage. When it comes in brackets, it stands for the matrix of all voltage variables; when it comes individually with index i, it stands for a single variable of voltage for bus i, and $V_{it,1}$ stands for the voltage parameter of bus i at period t for the fundamental frequency, $h = 1$. The notations are as follows:

Abbreviations		Ω^{PQ}	Set of all load buses
AHF	Active harmonic filter	Ω^{NL}	Set of all nonlinear buses
DER	Distributed energy resource	Ω^{NL}	Set of all nonlinear buses
DG	Distributed generation	Ω^{PQ}	Set of all load buses
DS	Distribution system	**Superscripts**	
GSC	Grid side converter	cha	Charge (for ESSs)
MG	Microgrid	dch	Discharge (for ESSs)
OC	Operation cost	spec	Harmonic spectrum
PCC	Point of common coupling	**Notations**	
PE	Power electronics	A	Branch current to bus injection incidence matrix
PQ	Power quality	B	Branch current to bus injection incidence matrix
SOC	State of charge (for ESSs)	c	SOC of ESS
THD	Total harmonic distortion	d	Distortion power
VSI	Voltage source inverter	G	Gain factor of GSC
Indices		i	Bus injection currents.
e	ESSs of the network	p	Active power.
f	Set of available AHF resources.	q	Reactive power.
h	Harmonic order.	S	Apparent power
i	Network buses.	u	Line currents
l	Network branches	v	Bus voltages
t	Operation horizon	y	Binary variable of ESS mode
Ω^{DG}	Set of all DG buses	$Z = R + jX$	Impedance = Resistance + Reactance
Ω^{ESS}	Set of all ESS buses	π	Price of energy

Appendix A

Data of the 33-bus IEEE test network are listed in Tables A1 and A2, respectively.

Table A1. Bus data of 33-bus test network (pu).

No	P_i	Q_i
1	0	0
2	0.022	0.013
3	0.02	0.009
4	0.027	0.018
5	0.013	0.007
6	0.013	0.004
7	0.044	0.022
8	0.044	0.022
9	0.013	0.004
10	0.013	0.004
11	0.01	0.007
12	0.013	0.008
13	0.013	0.008
14	0.027	0.018
15	0.013	0.002
16	0.013	0.004
17	0.013	0.004

Table A1. *Cont.*

No	P_i	Q_i
18	0.02	0.009
19	0.02	0.009
20	0.02	0.009
21	0.02	0.009
22	0.05	0.0225
23	0.02	0.011
24	0.013	0.004
25	0.093	0.044
26	0.013	0.006
27	0.093	0.044
28	0.013	0.004
29	0.027	0.016
30	0.044	0.133
31	0.033	0.016
32	0.047	0.022
33	0.013	0.009

Table A2. Line data of 33-bus network (pu).

No	From	To	R	X
1	1	2	0.0004	0.0002
2	2	3	0.002	0.001
3	3	4	0.0015	0.0008
4	4	5	0.0016	0.0008
5	5	6	0.0034	0.0029
6	6	7	0.0008	0.0026
7	7	8	0.0029	0.001
8	8	9	0.0042	0.003
9	9	10	0.0043	0.003
10	10	11	0.0008	0.0003
11	11	12	0.0015	0.0005
12	12	13	0.0061	0.0048
13	13	14	0.0022	0.0029
14	14	15	0.0024	0.0022
15	15	16	0.0031	0.0022
16	16	17	0.0053	0.0071
17	17	18	0.003	0.0024
18	2	19	0.0007	0.0006
19	19	20	0.0062	0.0056
20	20	21	0.0017	0.002
21	21	22	0.0029	0.0039
22	3	23	0.0019	0.0013
23	23	24	0.0037	0.0029
24	24	25	0.0037	0.0029
25	6	26	0.0008	0.0004
26	26	27	0.0012	0.0006
27	27	28	0.0044	0.0038
28	28	29	0.0033	0.0029
29	29	30	0.0021	0.0011
30	30	31	0.004	0.004
31	31	32	0.0013	0.0015
32	32	33	0.0014	0.0022

References

1. Koohi-Kamali, S.; Tyagi, V.; Rahim, N.; Panwar, N.; Mokhlis, H. Emergence of energy storage technologies as the solution for reliable operation of smart power systems: A review. *Renew. Sustain. Energy Rev.* **2013**, *25*, 135–165. [CrossRef]
2. Zakeri, B.; Syri, S. Electrical energy storage systems: A comparative life cycle cost analysis. *Renew. Sustain. Energy Rev.* **2015**, *42*, 569–596. [CrossRef]
3. Luo, X.; Wang, J.; Dooner, M.; Clarke, J. Overview of current development in electrical energy storage technologies and the application potential in power system operation. *Appl. Energy* **2015**, *137*, 511–536. [CrossRef]
4. Kondoh, J.; Ishii, I.; Yamaguchi, H.; Murata, A.; Otani, K.; Sakuta, K.; Higuchi, N.; Sekine, S.; Kamimoto, M. Electrical energy storage systems for energy networks. *Energy Convers. Manag.* **2000**, *41*, 1863–1874. [CrossRef]
5. Marini, A.; Ghazizadeh, M.-S.; Mortazavi, S.S.; Piegari, L. A harmonic power market framework for compensation management of DER based active power filters in microgrids. *Int. J. Electr. Power Energy Syst.* **2019**, *113*, 916–931. [CrossRef]
6. Beleiu, H.; Beleiu, I.; Pavel, S.; Darab, C. Management of Power Quality Issues from an Economic Point of View. *Sustainability* **2018**, *10*, 2326. [CrossRef]
7. Das, C.K.; Bass, O.; Kothapalli, G.; Mahmoud, T.S.; Habibi, D. Overview of energy storage systems in distribution networks: Placement, sizing, operation, and power quality. *Renew. Sustain. Energy Rev.* **2018**, *91*, 1205–1230. [CrossRef]
8. Buła, D.; Lewandowski, M. Comparison of frequency domain and time domain model of a distributed power supplying system with active power filters (APFs). *Appl. Math. Comput.* **2014**, *267*, 771–779. [CrossRef]
9. Kalair, A.; Abas, N.; Kalair, A.; Saleem, Z.; Khan, N. Review of harmonic analysis, modeling and mitigation techniques. *Renew. Sustain. Energy Rev.* **2017**, *78*, 1152–1187. [CrossRef]
10. Mohamad, F.; Teh, J.; Lai, C.-M.; Chen, L.-R. Development of energy storage systems for power network reliability: A review. *Energies* **2018**, *11*, 2278. [CrossRef]
11. Alshehri, J.; Khalid, M.; Alzahrani, A. An Intelligent Battery Energy Storage-Based Controller for Power Quality Improvement in Microgrids. *Energies* **2019**, *12*, 2112. [CrossRef]
12. Khani, H.; Farag, H.E. Optimal scheduling of energy storage to mitigate power quality issues in power systems. In Proceedings of the 2017 IEEE Power & Energy Society General Meeting, Chicago, IL, USA, 16–20 July 2017; pp. 1–5.
13. Ramos, G.; Rios, M.; Gómez, D.; Palacios, H.; Posada, L. Power quality study of large-scale wind farm with battery energy storage system. In Proceedings of the 2017 IEEE Industry Applications Society Annual Meeting, Cincinnati, OH, USA, 1–5 October 2017; pp. 1–6.
14. Yang, D. *Informatics in Control, Automation and Robotics*; Springer: Berlin/Heidelberg, Germany, 2012; Volume 2.
15. Arulampalam, A.; Barnes, M.; Jenkins, N.; Ekanayake, J.B. Power quality and stability improvement of a wind farm using STATCOM supported with hybrid battery energy storage. *IEE Proc. Gener. Transm. Distrib.* **2006**, *153*, 701–710. [CrossRef]
16. Marini, A.; Piegari, L.; Mortazavi, S.S.; Ghazizadeh, M.-S. Active Power Filter Commitment for Harmonic Compensation in Microgrids. In Proceedings of the 45th Annual Conference of the IEEE Industrial Electronics Society, IECON 2019, Lisbon, Portugal, 14–17 September 2019; pp. 7038–7044.
17. Das, J. *Power System Harmonics and Passive Filter Designs*; John Wiley & Sons: Hoboken, NJ, USA, 2015.
18. Abolhassani, M.T.; Enjeti, P.; Toliyat, H. Integrated doubly fed electric alternator/active filter (IDEA), a viable power quality solution, for wind energy conversion systems. *IEEE Trans. Energy Convers.* **2008**, *23*, 642–650. [CrossRef]
19. Abolhassani, M.T.; Enjeti, P.; Toliyat, H.A. Integrated doubly-fed electric alternator/active filter (IDEA), a viable power quality solution, for wind energy conversion systems. In Proceedings of the Conference Record of the 2004 IEEE Industry Applications Conference, 2004. 39th IAS Annual Meeting, Seattle, WA, USA, 3–7 October 2004; Volume 2033, pp. 2036–2043.
20. Ghatpande, O.A. Harmonic Compensation in A Grid Using Doubly Fed Induction Genertors. Master's Thesis, Missouri University of Science and Technology, Rolla, MO, USA, 2013.

21. Naidu, N.S.; Singh, B. Doubly fed induction generator for wind energy conversion systems with integrated active filter capabilities. *IEEE Trans. Ind. Inform.* **2015**, *11*, 923–933. [CrossRef]
22. Ying-Kwun, C. Determination of locations and sizes for active power line conditioners to reduce harmonics in power systems. *IEEE Trans. Power Deliv.* **1996**, *11*, 1610–1617. [CrossRef]
23. Hong, Y.-Y.; Hsu, Y.-L.; Chen, Y.-T. Three-phase active power line conditioner planning. In Generation, Transmission and Distribution, IEE Proceedings. *IET* **1998**, *145*, 281–287.
24. Alfonso, J.L.; Gonçalves, H.; Pinto, J. *Active Power Conditioners to Mitigate Power Quality Problems in Industrial Facilities*; IntechOpen Publisher: London, UK, 17 April 2013; Open Access.
25. Griffo, A.; Carpinelli, G.; Lauria, D.; Russo, A. An optimal control strategy for power quality enhancement in a competitive environment. *Int. J. Electr. Power Energy Syst.* **2007**, *29*, 514–525. [CrossRef]
26. Mortezaei, A.; Simoes, M.; Savaghebi, M.; Guerrero, J.; Al-Durra, A. Cooperative control of multi-master-slave islanded microgrid with power quality enhancement based on conservative power theory. *IEEE Trans. Smart Grid* **2016**, *99*, 1.
27. Wasiak, I.; Pawelek, R.; Mienski, R. Energy storage application in low-voltage microgrids for energy management and power quality improvement. *IET Gener. Transm. Distrib.* **2013**, *8*, 463–472. [CrossRef]
28. Cespedes, M.; Sun, J. Impedance modeling and analysis of grid-connected voltage-source converters. *IEEE Trans. Power Electron.* **2014**, *29*, 1254–1261. [CrossRef]
29. Cao, W.; Ma, Y.; Zhang, X.; Wang, F. Sequence impedance measurement of three-phase inverters using a parallel structure. In Proceedings of the Applied Power Electronics Conference and Exposition (APEC), Charlotte, NC, USA, 15–19 March 2015; pp. 3031–3038.
30. Cao, W. Impedance-Based Stability Analysis and Controller Design of Three-Phase Inverter-Based Ac Systems. Ph.D. Thesis, University of Tennessee, Knoxville, TN, USA, 2017.
31. Marini, A.; Mortazavi, S.; Piegari, L.; Ghazizadeh, M.-S. An efficient graph-based power flow algorithm for electrical distribution systems with a comprehensive modeling of distributed generations. *Electr. Power Syst. Res.* **2019**, *170*, 229–243. [CrossRef]
32. Farhoodnea, M.; Mohamed, A.; Shareef, H.; Zayandehroodi, H. Optimum placement of active power conditioner in distribution systems using improved discrete firefly algorithm for power quality enhancement. *Appl. Soft Comput.* **2014**, *23*, 249–258. [CrossRef]
33. Grady, W.M.; Santoso, S. Understanding power system harmonics. *IEEE Power Eng. Rev.* **2001**, *21*, 8–11. [CrossRef]
34. Marini, A.; Latify, M.A.; Ghazizadeh, M.S.; Salemnia, A. Long-term chronological load modeling in power system studies with energy storage systems. *Appl. Energy* **2015**, *156*, 436–448. [CrossRef]
35. Marini, A.; Latify, M.-A.; Ghazizadeh, M.-S.; Salemnia, A. Joint Maintenance Scheduling of Generation Units and Energy Storage Systems. *Tabriz J. Electr. Eng.* **2017**, *46*, 78.
36. Farahani, H.F.; Shayanfar, H.; Ghazizadeh, M. Modeling of stochastic behavior of plug-in hybrid electric vehicle in a reactive power market. *Electr. Eng.* **2014**, *96*, 1–13. [CrossRef]
37. *IEEE Recommended Practice for Monitoring Electric Power Quality*; IEEE Std 1159–2009 (Revis. IEEE Std 1159–1995); IEEE: Piscataway, NJ, USA, 2009; pp. 1–81. [CrossRef]
38. Abbas Marini, L.P.S.S.; Mortazavi, M.S. Ghazizadeh. In A linear programming approach to distribution Power Flow. In Proceedings of the 6th International Conference on Clean Electrical Power (ICCEP), Santa Margherita, Italy, 27–29 June 2017.
39. Barrows, C.; Bloom, A.; Ehlen, A.; Ikaheimo, J.; Jorgenson, J.; Krishnamurthy, D.; Lau, J.; McBennett, B.; O'Connell, M.; Preston, E. The IEEE reliability test system: A proposed 2019 Update. *IEEE Trans. Power Syst.* **2019**, *99*, 1. [CrossRef]
40. Rosenthal, E. GAMS-A user's guide. In Proceedings of the GAMS Development Corporation, Washington, DC, USA, 1–13 May 2008.

© 2020 by the authors. Licensee MDPI, Basel, Switzerland. This article is an open access article distributed under the terms and conditions of the Creative Commons Attribution (CC BY) license (http://creativecommons.org/licenses/by/4.0/).

MDPI
St. Alban-Anlage 66
4052 Basel
Switzerland
Tel. +41 61 683 77 34
Fax +41 61 302 89 18
www.mdpi.com

Energies Editorial Office
E-mail: energies@mdpi.com
www.mdpi.com/journal/energies

www.ingramcontent.com/pod-product-compliance
Lightning Source LLC
LaVergne TN
LVHW071955080526
838202LV00064B/6757